Springer Monographs in Mathematics

Goro Shimura

Elementary Dirichlet Series and Modular Forms

Springer

Goro Shimura
Department of Mathematics
Princeton University
Princeton, New Jersey 08544-1000
goro@math.princeton.edu

ISBN: 978-1-4419-2478-0 e-ISBN: 978-0-387-72474-4

Mathematics Subject Classification (2000): 11F11, 11F67, 11G05, 11G10, 11G15, 11R29, 11M06

Softcover reprint of the hardcover 1st edition 2007

Printed on acid-free paper.

9 8 7 6 5 4 3 2 1

springer.com

PREFACE

A book on any mathematical subject above textbook level is not of much value unless it contains new ideas and new perspectives. Also, the author may be encouraged to include new results, provided that they help the reader gain new insights and are presented along with known old results in a clear exposition.

It is with this philosophy that I write this volume. The two subjects, Dirichlet series and modular forms, are traditional, but I treat them in both orthodox and unorthodox ways. However, I try to make the book accessible to those who are not familiar with such topics, by including plenty of expository material. More specific descriptions of the contents will be given in the Introduction.

To some extent, this book has a supplementary nature to my previous book *Introduction to the Arithmetic Theory of Automorphic Functions,* published by Princeton University Press in 1971, though I do not write the present book with that intent. While the 1971 book grew out of my lectures in various places, the essential points of this new book have never been presented publicly or privately. I hope that it will draw an audience as large as that of the previous book.

Princeton
March 2007

Goro Shimura

TABLE OF CONTENTS

INTRODUCTION

There are two types of Dirichlet series that we discuss in this book:

(1) $$D^r_{a,N}(s) = \sum_{0 \neq n \in a+N\mathbf{Z}} n^r |n|^{-r-s},$$

(2) $$L^r(s;\, \alpha,\, \mathfrak{b}) = \sum_{0 \neq \xi \in \alpha+\mathfrak{b}} \xi^{-r} |\xi|^{r-2s}.$$

Here s is a complex variable as usual, r is 0 or 1 for $D^r_{a,N}$ and $0 < r \in \mathbf{Z}$ for the latter series; $a \in \mathbf{Z}$ and $0 < N \in \mathbf{Z}$. To define the series of (2) we take an imaginary quadratic field K embedded in $\mathbf{C}$ and take also an element α of K and a $\mathbf{Z}$-lattice $\mathfrak{b}$ in K. One of our principal problems is to investigate the nature of the values of these series at certain integer values of s. As a preliminary step, we discuss their analytic continuation and functional equaltions. We obtain Dirichlet L-functions and certain Hecke L-functions of K as suitable linear combinations of these series, and so the values of such L-functions are included in our objects of study. As will be explained below, these series are directly and indirectly related to elliptic modular forms, and the exposition of such functions in that context forms a substantial portion of this volume. Thus, as we said in the preface, the main objective of this book is to present some new ideas, new results, and new perspectives, along with old ones in this area covering certain aspects of the theory of modular forms and Dirichlet series.

To be more specific, let us first consider the Dirichlet L-function

$$L(s,\, \chi) = \sum_{n=1}^{\infty} \chi(n) n^{-s}$$

with a primitive Dirichlet character χ modulo a positive integer N. It is well known that if k is a positive integer and $\chi(-1) = (-1)^k$, then

(3) $$\frac{2k!G(\overline{\chi})}{(2\pi i)^k} L(k,\, \chi) = -\sum_{a=1}^{N} \overline{\chi}(a) B_k(a/N),$$

where B_k is the Bernoulli polynomial of degree k and $G(\overline{\chi})$ is the Gauss sum of $\overline{\chi}$. If $k = 1$ in particular, the last sum for χ such that $\chi(-1) = -1$ becomes $\sum_{a=1}^{N} \overline{\chi}(a)a/N$, which reminds us of another well known result about the second factor of the class number of a cyclotomic field, which is written h_K/h_F. Here

h_K resp. h_F is the class number of K resp. F; K is an imaginary subfield of $\mathbf{Q}(\zeta)$ with a primitive mth root of unity ζ for some positive integer m and F is the maximal real subfield of K. There is a classical formula for h_K/h_F, which is easy factors times a product $\prod_{\chi\in X}\sum_a \chi(a)a$, where X is a certain set of primitive Dirichlet characters χ such that $\chi(-1)=-1$.

No alternative formulas have previously been presented except when K is an imaginary quadratic field $K=\mathbf{Q}(\sqrt{-d})$. Given such a K, take a real character χ of conductor d that corresponds to K. Then it is well known that

$$(4)\qquad \frac{w_K\sqrt{d}}{2\pi}L(1,\,\chi)=h_K=\frac{w_K}{2(2-\chi(2))}\sum_{a=1}^{q}\chi(a),\quad q=[(d-1)/2],$$

where w_K is the number of roots of unity in K.

Now we will prove as one of the main results of this book that there are new formulas for $L(k,\,\chi)$. The most basic one is

$$(5)\qquad \frac{(k-1)!G(\overline{\chi})}{(2\pi i)^k}L(k,\,\chi)=\frac{1}{2(2^k-\chi(2))}\sum_{a=1}^{q}\overline{\chi}(a)E_{1,k-1}(2a/d),$$

where $q=[(d-1)/2]$ and $E_{1,k-1}(t)$ is the Euler polynomial of degree $k-1$. This clearly includes (4) (without h_K) as a special case, as $E_{1,0}(t)=1$. This formula is better than (3) at least from the computational viewpoint, as $E_{1,k-1}(t)$ is a polynomial in t of degree $k-1$, whereas $B_k(t)$ is of degree k. We will present many more new formulas for $L(k,\,\chi)$ in Sections 4 and 6. As applications, we will prove some new formulas for the quotient h_K/h_F.

To avoid excessive details, we state it in this introduction only when $K=\mathbf{Q}(\zeta)$ with $m=2^r>4$:

$$\frac{h_K}{h_F}=2^{\gamma}\prod_{s=3}^{r}\prod_{\chi\in Y_s}\left\{\sum_{a=1}^{d_s}\chi(a)\right\},\quad d_s=2^{s-2}-1,\quad \gamma=r-1-2^{r-2}.$$

Here Y_s is the set of primitive Dirichlet characters χ of conductor 2^s such that $\chi(-1)=-1$. Notice that we have $\sum_{a=1}^{d_s}\chi(a)$, which is of far "smaller size" than the sum $\sum_a \chi(a)a$ in the classical formula. A similar but somewhat different formula can be obtained in the case $m=\ell^r$ with an odd prime ℓ.

The latter part of the book concerns the critical values of the series of type (2). In this case we evaluate it at $k/2$ with an integer k such that $2-r\le k\le r$ and $r-k\in 2\mathbf{Z}$. Then we can show that:

(6) *There is a constant γ which depends only on K (that is, independent of α, $\mathfrak{b}$, r, and k) such that $L^r(k/2;\,\alpha,\,\mathfrak{b})$ is $\pi^{(r+k)/2}\gamma^r$ times an algebraic number.*

This was proved in one of the author's papers. Though we will give a proof of this fact in this book, it is merely the starting point. Indeed one of our main problems is to find a suitable γ so that the algebraic number can be computed. Before going into this problem, we note that the constant γ can be given as $\varphi(\tau)$ with a modular form φ of weight 1 and $\tau \in K \cap H$, where $H = \{z \in \mathbf{C} \mid \mathrm{Im}(z) > 0\}$.

If $r = k \neq 2$, the value $L^r(r/2;\, \alpha,\, \mathfrak{b})$ can be given as $\pi^r h(\tau)$ with a holomorphic modular form h of weight r. Thus our task is to find $(h/\varphi^r)(\tau)$. This can be achieved as follows. We fix a congruence subgroup Γ of $SL_2(\mathbf{Z})$ to which both φ and h belong, and assume that we can find two modular forms f and g that generate the algebra of all modular forms of all nonnegative weights with respect to Γ. Then $h = P(f, g)$ with a polynomial P, and $(h/\varphi^r)(\tau) = P\big((f/\varphi^k)(\tau),\, (g/\varphi^\ell)(\tau)\big)$, where k resp. ℓ is the weight of f resp. g. However there are two essential questions:

(I) How can we find P?

(II) In the general case in which $r \neq k$ or $k = 2$, $L^r(k/2;\, \alpha,\, \mathfrak{b})$ can be given as $\pi^{(r+k)/2}p(\tau)$ with some nonholomorphic modular form (which we call *nearly holomorphic*) p of weight r. Then, how can we handle $(p/\varphi^r)(\tau)$?

Problem (II) can be reduced to Problem (I) and Problem (II) for simpler p. In the easiest case we can express p as a polynomial $\sum_{a=0}^{[r/2]} E_2^a h_a$, where h_a is a holomorphic modular form of weight $r - 2a$ and E_2 is a well known nonholomorphic Eisenstein series of weight 2:

$$E_2(z) = \frac{1}{8\pi y} - \frac{1}{24} + \sum_{n=1}^{\infty}\Bigg(\sum_{0<d|n} d\Bigg)\mathbf{e}(nz).$$

Then the problem can be reduced to $(E_2/\varphi^2)(\tau)$ and $(h_a/\varphi^{r-2a})(\tau)$. The latter quantity is handled by P for h_a. As for E_2, we have to deal with it in a special way. For a given τ we will find a special holomorphic modular form q of weight 2 such that $(E_2/q)(\tau)$ can be explicitly given.

In this way the value of nonholomorphic functions can be reduced to the case of holomorphic modular forms, and to Problem (I). We also have to find $(f/\varphi^k)(\tau)$ and $(g/\varphi^\ell)(\tau)$, which is nontrivial, but our idea is to reduce infinitely many values to finitely many values. In general, there is no clear-cut answer to (I). However, we can produce two types of recurrence formulas for Eisenstein series, which seem to be new and by which the problem about h/φ^r of an arbitrary weight r can be reduced to the case of smaller r. (See (10.8) and (10.15c).) Without stating it, we content ourselves by mentioning its application to $L^r(k/2;\, \alpha,\, \mathfrak{b})$. Assuming that $\alpha \notin \mathfrak{b}$, $0 < k \leq r$, and $r - k \in 2\mathbf{Z}$, put $n = (r-k)/2$, and

$$\mathfrak{L}_k^n(\alpha,\, \mathfrak{b}) = (-1)^k(2\pi i)^{-k-n}\Gamma(k+n)L^r(k/2;\, \alpha,\, \mathfrak{b}) - \begin{cases} 2(2i)^n\mathrm{Im}(\tau)^n(D_2^n E_2)(\tau) & \text{if } k = 2, \\ 0 & \text{if } k \neq 2, \end{cases}$$

where D_2^n is a differential operator of the type mentioned above. Then we have a recurrence formula

$$\mathfrak{L}_{t+5}^n(\alpha,\, \mathfrak{b}) = 12\sum_{i=0}^{t}\binom{t}{i}\sum_{j=0}^{n}\binom{n}{j}\,\mathfrak{L}_{i+3}^{j}(\alpha,\, \mathfrak{b})\cdot\mathfrak{L}_{t-i+2}^{n-j}(\alpha,\, \mathfrak{b})$$

for $0 \le t \in \mathbf{Z}$ and $0 \le n \in \mathbf{Z}$. Thus the values of $\mathfrak{L}_k^n(\alpha,\, \mathfrak{b})$ for $k > 4$ can be reduced inductively to those for $2 \le k \le 4$. If $\alpha \in \mathfrak{b}$, there is another recurrence formula which reduces $\mathfrak{L}_{2k}^n(\alpha,\, \mathfrak{b})$ for $2k \ge 8$ to the cases with $2k = 4$ and $2k = 6$.

We can form a Hecke L-function $L(s,\, \lambda) = \sum_{\mathfrak{a}} \lambda(\mathfrak{a})N(\mathfrak{a})^{-s}$, with a Hecke ideal character λ of K such that

$$\lambda(\alpha\mathfrak{r}) = \alpha^{-r}|\alpha|^r \quad \text{if} \quad \alpha \in K^\times \quad \text{and} \quad \alpha - 1 \in \mathfrak{c},$$

where $\mathfrak{c}$ is an integral ideal of K. Since this is a finite linear combination of series of type (2), statement (6) holds for $L(k/2,\, \lambda)$ in place of $L^r(k/2;\, \alpha,\, \mathfrak{b})$. In Section 13, we will present many examples of numerical values of $L(k/2,\, \lambda)$.

When $r = 1$, the function $L(s,\, \lambda)$ is closely connected with an elliptic curve C defined over an algebraic number field h with complex multiplication in K. In a certain case, it is indeed the zeta function of C over h. We will study this aspect in Section 14, and compare $L(1/2,\, \lambda)$ with a period of a holomorphic 1-form on C, when C is a member of a one-parameter family $\{C_z\}_{z \in H}$ of elliptic curves.

However, without going into details of this theory, let us end this introduction by briefly mentioning some other noteworthy features of the book.

(A) A discussion of irregular cusps of a congruence subgroup of $SL_2(\mathbf{Z})$ in §1.11 and Theorem 1.13.

(B) The functional equation of the Eisenstein series

$$\mathfrak{E}_k^N(z,\, s;\, p,\, q) = \mathrm{Im}(z)^s \sum_{(m,\, n)} (mz+n)^{-k}|mz+n|^{-2s}$$

under $s \mapsto 1-k-s$ (Theorem 9.7). Here $(z,\, s) \in H \times \mathbf{C}$, $0 < N < \mathbf{Z}$, $0 \le k \in \mathbf{Z}$, $(p,\, q) \in \mathbf{Z}^2$, and $(m,\, n)$ runs over $\mathbf{Z}^2$ under the condition $0 \ne (m,\, n) \equiv (p,\, q) \pmod{N\mathbf{Z}^2}$.

(C) The explicit Fourier expansion of $\mathfrak{E}_k^N(z,\, 1-k;\, p,\, q)$ given in (9.14).

(D) In Section 15, we discuss isomorphism classes of abelian varieties, elliptic curves in particular, with complex multiplication defined over a number field with the same zeta function.

(E) In Section 16 we present a new class of holomorphic differential operators $\{\mathfrak{A}_k^p\}_{p=2}^{\infty}$. The operator $\mathfrak{A}_k^p$ sends an automorphic form of weight k to that of weight $kp + 2p$, and every operator of the same nature can be reduced to this class.

CHAPTER I

PRELIMINARIES ON MODULAR FORMS AND DIRICHLET SERIES

1. Basic symbols and the definition of modular forms

Though some basic facts on elliptic modular forms are reviewed in this section, we do not need them in Sections 2 through 7. Therefore the reader may go directly to Section 2 after reading §1.1, Lemmas 1.6 and 1.12, and return to this section before going to Section 8.

1.1. The symbols **Z**, **Q**, **R**, and **C** will mean as usual the ring of integers, the fields of rational numbers, real numbers, and complex numbers, respectively. Also, we denote by $\overline{\mathbf{Q}}$ the algebraic closure of **Q** in **C**. Given an associative ring A with identity element, we denote by $A^\times$ the group of all invertible elements of A, and by $M_n(A)$ the ring of all $n \times n$-matrices with entries in A, and put $GL_n(R) = M_n(A)^\times$. The identity element of $M_n(A)$ is denoted by 1_n, or simply by 1, and the transpose of a matrix X by ${}^tX$. If A is commutative, we put

$$SL_n(A) = \{\alpha \in GL_n(A) \mid \det(\alpha) = 1\}.$$

Given a (2×2)-matrix γ with coefficients in any ring, we put $\gamma = \begin{bmatrix} a_\gamma & b_\gamma \\ c_\gamma & d_\gamma \end{bmatrix}$ whenever no confusion is expected. We now put

$$H = \{\, z \in \mathbf{C} \mid \mathrm{Im}(z) > 0\,\}, \tag{1.1a}$$

$$G = GL_2(\mathbf{Q}), \qquad G^1 = SL_2(\mathbf{Q}), \tag{1.1b}$$

$$G_{\mathbf{a}} = GL_2(\mathbf{R}), \qquad G^1_{\mathbf{a}} = SL_2(\mathbf{R}), \tag{1.1c}$$

$$G_{\mathbf{a}+} = \{\alpha \in GL_n(\mathbf{R}) \mid \det(\alpha) > 0\}, \qquad G_+ = G \cap G_{\mathbf{a}+}. \tag{1.1d}$$

For $\gamma \in G_{\mathbf{a}+}$ and $z \in H$ we define $\gamma(z) \in H$ and $j_\gamma(z) \in \mathbf{C}^\times$ by

$$\gamma(z) = \gamma z = (a_\gamma z + b_\gamma)/(c_\gamma z + d_\gamma), \tag{1.2a}$$

$$j(\gamma, z) = j_\gamma(z) = \det(\gamma)^{-1/2}(c_\gamma z + d_\gamma). \tag{1.2b}$$

These can be expressed by a single equality

$$\det(\gamma)^{-1/2}\gamma \begin{bmatrix} z \\ 1 \end{bmatrix} = \begin{bmatrix} \gamma(z) \\ 1 \end{bmatrix} j_\gamma(z). \tag{1.3}$$

We recall easy relations

(1.4a) $$j_{\alpha\beta}(z) = j_\alpha(\beta z) j_\beta(z),$$

(1.4b) $$\mathrm{Im}(\alpha z) = |j_\alpha(z)|^{-2}\mathrm{Im}(z), \qquad d(\alpha z)/dz = j_\alpha(z)^{-2}.$$

In fact, we can define γz by (1.2a) even for $\gamma \in GL_2(\mathbf{C})$ and $z \in \mathbf{C} \cup \{\infty\}$. Then the last formula of (1.4b) is meaningful for $\alpha \in GL_2(\mathbf{C})$.

1.2. For a function $f : H \to \mathbf{C}$, $k \in \mathbf{Z}$, and $\alpha \in G_{\mathbf{a}+}$, we define $f\|_k\alpha : H \to \mathbf{C}$ by

(1.5) $$(f\|_k\alpha)(z) = j_\alpha(z)^{-k} f(\alpha(z)) \qquad (z \in H).$$

We have

(1.6a) $$f\|_k(\alpha\beta) = (f\|_k\alpha)\|_k\beta,$$

(1.6b) $$f\|_k(c1_2) = \mathrm{sgn}(c)^k f \quad \text{if} \quad c \in \mathbf{R}^\times.$$

For a positive integer N we put

(1.7a) $$\Gamma(N) = \{\, \gamma \in SL_2(\mathbf{Z}) \mid \gamma \equiv 1_2 \pmod{N} \,\},$$

(1.7b) $$\Gamma^0(N) = \{\, \gamma \in SL_2(\mathbf{Z}) \mid b_\gamma \in N\mathbf{Z} \,\},$$

(1.7c) $$\Gamma_0(N) = \{\, \gamma \in SL_2(\mathbf{Z}) \mid c_\gamma \in N\mathbf{Z} \,\},$$

(1.7d) $$\Gamma_1(N) = \{\, \gamma \in \Gamma_0(N) \mid a_\gamma - 1 \in N\mathbf{Z} \,\}.$$

Then $\Gamma(1) = SL_2(\mathbf{Z})$, $\Gamma(N) \subset \Gamma_1(N) \subset \Gamma_0(N)$, and $\Gamma(N)$ is a normal subgroup of $\Gamma(1)$. We call a sugroup of $\Gamma(1)$ a **congruence subgroup** if it contains $\Gamma(N)$ as a subgroup of finite index for some N.

1.3. Let us now recall the definition of a modular form. We refer the reader to [S71] for the basic facts on this subsect. We first put, for $c \in \mathbf{C}$,

(1.8) $$\mathbf{e}(c) = \exp(2\pi i c).$$

Given a congruence subgroup Γ and an integer k, we call a holomorphic function f on H a (holomorphic) **modular form of weight k with respect to Γ** if the following two conditions are satisfied:

(1.9a) $f\|_k\gamma = f$ *for every* $\gamma \in \Gamma$.

(1.9b) *For every* $\alpha \in \Gamma(1)$ *one has* $(f\|_k\alpha)(z) = \sum_{n=0}^{\infty} c_{\alpha n} \cdot \mathbf{e}(nz/N_\alpha)$ *with* $c_{\alpha n} \in \mathbf{C}$ *and* $0 < N_\alpha \in \mathbf{Z}$.

We denote by $\mathscr{M}_k(\Gamma)$ the set of all such f. The last condition implies in particular

(1.10) $$f(z) = \sum_{n=0}^{\infty} c_n \cdot \mathbf{e}(nz/N)$$

with $c_n \in \mathbf{C}$ and $0 < N \in \mathbf{Z}$. It is known that: (i) $\mathscr{M}_k(\Gamma)$ is a complex vector space of finite dimension; (ii) $\mathscr{M}_k(\Gamma) = \{0\}$ if $k < 0$; (iii) $\mathscr{M}_0(\Gamma) = \mathbf{C}$. From (1.6b) we see that $\mathscr{M}_k(\Gamma) = \{0\}$ if k is odd and $-1 \in \Gamma$. It is often convenient to consider modular forms without referring to Γ, so we put

(1.11) $$\mathscr{M}_k = \bigcup_\Gamma \mathscr{M}_k(\Gamma),$$

where Γ runs over all congruence subgroups of $\Gamma(1)$. We call an element f of $\mathscr{M}_k$ a **cusp form** if $c_{\alpha 0}$ of (1.9b) is 0 for every $\alpha \in \Gamma(1)$. We denote by $\mathscr{S}_k$ the subset of $\mathscr{M}_k$ consisting of all the cusp forms, and put $\mathscr{S}_k(\Gamma) = \mathscr{S}_k \cap \mathscr{M}_k(\Gamma)$. For example, we recall a classical fact that $\mathscr{S}_{12}(\Gamma(1)) = \mathbf{C}\Delta$ with

(1.12) $$\Delta(z) = \mathbf{q}\prod_{n=1}^{\infty}(1-\mathbf{q}^n)^{24}, \qquad \mathbf{q} = \mathbf{e}(z).$$

Moreover, for $r \in \mathbf{Z}$ the function $\Delta^{r/24}$ can be defined by

(1.13) $$\Delta^{r/24}(z) = \mathbf{e}(rz/24)\prod_{n=1}^{\infty}(1-\mathbf{q}^n)^r,$$

and $\Delta^{r/24}(z) \in \mathscr{S}_{r/2}$ if $0 < r \in 2\mathbf{Z}$. These functions are nonzero everywhere on H. Let us now put

(1.14) $$P_+ = \{\alpha \in G_+ \mid c_\alpha = 0\}.$$

Clearly $P_+ = \{\alpha \in G_+ \mid \alpha(\infty) = \infty\}$. We have

(1.15) $$G_+ = \Gamma(1)P_+.$$

Indeed, if $\alpha \in G_+$ and $\alpha(\infty) \neq \infty$, then we can put $\alpha(\infty) = a/c$ with integers a and c that are relatively prime. We can find integers b and d such that $ad - bc = 1$. Put $\gamma = \begin{bmatrix} a & b \\ c & d \end{bmatrix}$. Then $\gamma \in \Gamma(1)$ and $\gamma(\infty) = a/c = \alpha(\infty)$, and so $\gamma^{-1}\alpha \in P_+$. This proves (1.15). Because of this equality, we can replace $\Gamma(1)$ in condition (1.9b) by G_+.

1.4. Given a subfield Φ of $\mathbf{C}$, we denote by $\mathscr{M}_k(\Phi, \Gamma)$ the set of all elements f of $\mathscr{M}_k(\Gamma)$ of the form (1.10) with $c_n \in \Phi$ for all n. We then put $\mathscr{S}_k(\Phi, \Gamma) = \mathscr{S}_k \cap \mathscr{M}_k(\Phi, \Gamma)$. Furthermore, we put

$$\mathscr{M}_k(\Phi) = \bigcup_\Gamma \mathscr{M}_k(\Phi, \Gamma), \qquad \mathscr{S}_k(\Phi) = \bigcup_\Gamma \mathscr{S}_k(\Phi, \Gamma),$$

where Γ runs over all congruence subgroups of $\Gamma(1)$.

We extend this to meromorphic functions as follows. For $m \in \mathbf{Z}$ and Φ as above, we denote by $\mathscr{A}_m(\Phi)$ the set of all quotients p/q such that $p \in \mathscr{M}_{k+m}(\Phi)$ and $0 \neq q \in \mathscr{M}_k(\Phi)$ with any $k \in \mathbf{Z}, > 0$. We then put $\mathscr{A}_m = \mathscr{A}_m(\mathbf{C})$,

$$\mathscr{A}_m(\Gamma) = \{f \in \mathscr{A}_m \mid f\|_m\gamma = f \text{ for every } \gamma \in \Gamma\},$$
$$\mathscr{A}_m(\Phi, \Gamma) = \mathscr{A}_m(\Phi) \cap \mathscr{A}_m(\Gamma).$$

We call the elements of $\mathscr{A}_m(\Phi)$ **Φ-rational.** The elements of $\mathscr{A}_0(\Gamma)$ are called **modular functions** with respect to Γ. The orbit space $\Gamma\backslash(H \cup \mathbf{Q} \cup \{\infty\})$ has a structure of a compact Riemann surface, which can be presented as an algebraic curve defined over an algebraic number field. Thus $\mathscr{A}_0(\Phi, \Gamma)$ can be identified with the field of all algebraic functions on that curve over Φ for a suitable choice of Φ. For these the reader is referred to [S71, §6.7].

Let $\mathbf{Q}_{ab}$ denote the maximal abelian extension of $\mathbf{Q}$ in $\mathbf{C}$. Then the field $\mathscr{A}_0(\mathbf{Q}_{ab})$ is stable under the maps $f \mapsto f \circ \alpha$ for all $\alpha \in G_+$. To show this, by (1.15) we can reduce the problem to the cases where α belongs to $\Gamma(1)$ or P_+. The case $\alpha \in P_+$ is obvious. If $\alpha \in \Gamma(1)$, the result can be derived from the fact that $\mathscr{A}_0((\Gamma(N))$ is generated by $J(z)$ and "modified" N-division values of the Weierstrass $\wp$-function as stated in [S71, Proposition 6.9]. In [S71] the field $\mathscr{A}_0(\mathbf{Q}_{ab})$ is denoted by $\mathfrak{F}$, and the stability of $\mathfrak{F}$ under G_+ is given in [S71, Proposition 6.22]. Somewhat more strongly we have

Theorem 1.5. *Let Φ be a subfield of $\mathbf{C}$ containing $\mathbf{Q}_{ab}$. Then $f\|_m\alpha \in \mathscr{A}_m(\Phi)$ for every $f \in \mathscr{A}_m(\Phi)$ and every $\alpha \in G_+$. In particular, $f\|_m\alpha \in \mathscr{M}_m(\Phi)$ for every $\alpha \in G_+$ if $f \in \mathscr{M}_m(\Phi)$,*

PROOF. The field $\mathscr{A}_0$ is the composite of $\mathbf{C}$ and $\mathscr{A}_0(\mathbf{Q}_{ab})$; also, $\mathscr{A}_0(\mathbf{Q}_{ab})$ and $\mathbf{C}$ are linearly disjoint over $\mathbf{Q}_{ab}$; see [S71, Proposition 6.1, Theorem 6.6 (4), Proposition 6.27]. Therefore $\mathscr{A}_0(\Phi)$ is the composite of Φ and $\mathscr{A}_0(\mathbf{Q}_{ab})$, and so it is stable under G_+. Since $G_+ = \Gamma(1)P_+$ and our theorem is clear for $\alpha \in P_+$, it is sufficient to prove the case $\alpha \in \Gamma(1)$. Given $f \in \mathscr{A}_m(\Phi)$, we put $h = f/g$ with $g = \Delta^{m/12}$. From (1.13) we see that $g \in \mathscr{A}_m(\mathbf{Q})$, and so $h \in \mathscr{A}_0(\Phi)$. Thus $h \circ \alpha \in \mathscr{A}_0(\Phi)$ for every $\alpha \in \Gamma(1)$ for the reason explained above. Since $\Delta\|_{12}\alpha = \Delta$, we see that $g\|_m\alpha$ is g times a twelfth root of unity, and so it is contained in $\mathscr{A}_m(\Phi)$. This proves the first assertion of our theorem, as $f\|_m\alpha = (h \circ \alpha)g\|_m\alpha$, from which the assertion for $f \in \mathscr{M}_m(\Phi)$ follows immediately.

Lemma 1.6. *For a function f on H given by $f(z) = \sum_{n=0}^{\infty} a_n \mathbf{e}(nz/N)$ with $a_n \in \mathbf{C}$ and $0 < N \in \mathbf{Z}$, the following assertions hold. (In each statement, α is a positive constant.)*

(i) $f(x+iy) - a_0 = O(e^{-2\pi y/N})$ *uniformly as* $y \to \infty$.

(ii) $a_n = O(n^\alpha) \implies f(x+iy) = O(y^{-\alpha-1})$ *uniformly as* $y \to 0$.

(iii) $\sum_{n=0}^{\infty} |a_n| n^{-\alpha} < \infty \implies f(x+iy) = O(y^{-\alpha})$ *uniformly as* $y \to 0$.

(iv) $f(x+iy) = O(y^{-\alpha})$ *uniformly as* $y \to 0 \implies a_n = O(n^\alpha)$.

(v) $\overline{f(z)} = f(-\overline{z})$ *if* $a_n \in \mathbf{R}$ *for all* n.

PROOF. Changing f for $f(Nz)$, we may assume that $N = 1$. Assertion (i) is clear, as f is a convergent power series in $e^{2\pi iz}$. We will prove (ii) in the next section after (2.14). To prove (iii), let $g = f - a_0$. Then $\sum_{k=1}^{n} |a_k| \le \sum_{k=1}^{n} |a_k|(n/k)^\alpha \le n^\alpha \sum_{k=1}^{n} |a_k| k^{-\alpha} \le Bn^\alpha$ with some $B > 0$. Thus $(1 - e^{-2\pi y})^{-1}|g(z)| \le \sum_{m=0}^{\infty} e^{-2\pi my} \cdot \sum_{n=1}^{\infty} |a_n| e^{-2\pi ny} = \sum_{n=1}^{\infty} e^{-2\pi ny} \sum_{k=1}^{n} |a_k| \le B\sum_{n=1}^{\infty} n^\alpha e^{-2\pi ny} = O(y^{-\alpha-1})$ as in (ii). Therefore we obtain (iii), as $1-e^{-2\pi y} = O(y)$ as $y \to 0$. For f as in (iv) we have $|a_n e^{-2\pi ny}| = \left|\int_0^1 f(x+iy)e^{-2\pi inx}\,dx\right| \le$

$Ay^{-\alpha}$ with some $A > 0$ for sufficiently small y. Taking $y = 1/n$, we obtain (iv). The last assertion is an easy exercise.

Lemma 1.7. *Given $f \in \mathscr{M}_k$, there exists a positive constant K such that $|f(z)| \le K(1 + y^{-k})$ on the whole H. If $f \in \mathscr{S}_k$, then we can take K so that $|f(z)| \le Ky^{-k/2}$ on the whole H.*

PROOF. Suppose $f \in \mathscr{M}_k(\Gamma)$. Let $T = \{z \in H \mid \mathrm{Im}(z) > 1/2\}$. Since T contains the well known fundamental domian for $\Gamma(1)\backslash H$, we have $H = \Gamma(1)T$, and so we can find a finite subset A of $\Gamma(1)$ such that $H = \bigcup_{\alpha\in A} \Gamma\alpha T$. Put $f_\alpha = f\|_k\alpha$. Given $z \in H$, there exists $\alpha \in A$ such that $z \in \gamma\alpha T$ with some $\gamma \in \Gamma$. Put $\beta = \alpha^{-1}\gamma^{-1}$. Then $f = f\|_k\alpha\beta = f_\alpha\|_k\beta = j_\beta(z)^{-k} f_\alpha(\beta z)$. Now f_α is bounded on T, and hence there is a positive constant K such that $|f_\alpha(w)| \le K$ for every $w \in T$ and every $\alpha \in A$. Thus $|f(z)| \le K|c_\beta z + d_\beta|^{-k}$, as $\beta z \in T$. If $c_\beta \ne 0$, then $|f(z)| \le K|c_\beta y|^{-k} \le Ky^{-k}$, as $|c_\beta| \ge 1$; if $c_\beta = 0$, then $|f(z)| \le K$. This proves the first assertion. Suppose that f is a cusp form. Put $g(z) = y^k|f|^2$ and observe that g is Γ-invariant and $g\circ\alpha = y^k|f_\alpha|^2$, which is bounded on T, as f_α is a cusp form; see Lemma 1.6 (i). Thus we can find a positive constant M such that $|g(\alpha z)| \le M$ for every $\alpha \in A$ and $z \in T$. Given $z \in H$, take $\gamma \in \Gamma$, $\alpha \in A$, and $w \in T$ so that $z = \gamma\alpha w$. Then $|y^k f(z)^2| = g(z) = g(\gamma\alpha w) = g(\alpha w) \le M$. This completes the proof.

Lemma 1.8. *If $f(z) = \sum_{n=0}^{\infty} a_n \mathbf{e}(nz/N) \in \mathscr{M}_k$, then $a_n = O(n^k)$. If in particular f is a cusp form, then $a_n = O(n^{k/2})$.*

This follows from Lemma 1.6 (iv) combined with Lemma 1.7.

Theorem 1.9. (i) *For every $k \in \mathbf{Z}$, > 0, we have $\mathscr{M}_k = \mathscr{M}_k(\mathbf{Q}) \otimes_{\mathbf{Q}} \mathbf{C}$.*

(ii) *Given $f(z) = \sum_{n=0}^{\infty} a_n \mathbf{e}(nz/N) \in \mathscr{M}_k$ and a field-automorphism σ of* $\mathbf{C}$, *define an infinite series $f^\sigma(z)$ formally by $f^\sigma(z) = \sum_{n=0}^{\infty} a_n^\sigma \mathbf{e}(nz/N)$. Then this defines a holomorphic function on H and $f^\sigma \in \mathscr{M}_k$.*

PROOF. Assertion (i) can easily be derived from the facts that the curve $C = \Gamma(N)\backslash(H \cup \mathbf{Q} \cup \{\infty\})$ has a $\mathbf{Q}$-rational model and that for every $\mathbf{Q}$-rational divisor X on C the linear system $\mathscr{L}(X)$ has a basis contained in $\mathscr{A}_0(\mathbf{Q}, \Gamma(N))$. These are explained in [S00, pp. 62–64] in a more general case; (i) is actually a special case of [S00, Theorem 9.9]. Given f as in (ii), we can put, in view of (i), $f = \sum_{g\in X} c_g g$ with a finite subset X of $\mathscr{M}_k(\mathbf{Q})$ and $c_g \in \mathbf{C}$. Then clearly $f^\sigma = \sum_{g\in X} c_g^\sigma g$, which proves (ii).

We can also prove that $\mathscr{M}_k(\mathbf{Q})$ can be spanned by the elements whose Fourier coefficients at ∞ are contained in $\mathbf{Z}$; see [S75d, Theorem 1] and [S76a, pp. 682–683]. See also [S71, Theorem 3.32], though the result stated there concerns only cusp forms. We will present explicit examples of generators of $\sum_{k=0}^{\infty} \mathscr{M}_k(\mathbf{Q})$ in Section 10.

1.10. In this and next subsections we recall some basic terms such as elliptic points and cusps, and discuss how they are related to the dimension formula for $\mathscr{M}_k(\Gamma)$. The full treatment of these topics can be found in the first two chapters of [S71]. First of all, we call an element α of G^1 **elliptic** if $\alpha \neq \pm 1$ and α has a fixed point on H. Such a fixed point is unique for α. Let Γ be a congruence subgroup of $\Gamma(1)$. An element α, $\neq \pm 1$, of Γ is elliptic if and only if α is of finite order. The order of an elliptic element of Γ is 3, 4, or 6. By an **elliptic point** of Γ we understand a point fixed by an elliptic element of Γ. The images of an elliptic point of Γ under Γ are also elliptic points of Γ. We can then find a finite complete set of representatives for the elliptic points of Γ modulo Γ. Given an elliptic point w of Γ, we put

$$\overline{\Gamma}_w = \{\gamma \in \{\pm 1\}\Gamma \mid \gamma(w) = w\}. \tag{1.16}$$

Then $\overline{\Gamma}_w/\{\pm 1\}$ is of order 2 or 3. We call w an elliptic point of Γ of **order** 2 or 3 accordingly.

Next, there is a notion of a **cusp.** Since we are considering a subgroup Γ of $\Gamma(1)$, the set of cusps of Γ is merely $\mathbf{Q} \cup \{\infty\}$. Put

$$P^1 = \{\alpha \in G^1 \mid c_\alpha = 0\}, \qquad \Gamma_P = \Gamma \cap P^1. \tag{1.17}$$

In view of (1.15) we have $G^1 = \Gamma(1)P_1$, and the map $\alpha \mapsto \alpha(\infty)$ gives a bijection of $\Gamma(1)/\Gamma(1)_P$ onto $\mathbf{Q} \cup \{\infty\}$. Thus $\Gamma \backslash (\mathbf{Q} \cup \{\infty\})$ can be identified with $\Gamma \backslash \Gamma(1)/\Gamma(1)_P$, which is clearly a finite set. For $s \in \mathbf{Q} \cup \{\infty\}$ put

$$\Gamma_s = \{\alpha \in \Gamma \mid \alpha(s) = s\} \tag{1.18}$$

and let ρ be an element of $\Gamma(1)$ such that $\rho(s) = \infty$. Then we can find a positive integer h such that

$$\{\pm 1\}\rho\Gamma_s\rho^{-1} = \{\pm 1\}\left\{ \begin{bmatrix} 1 & h \\ 0 & 1 \end{bmatrix}^n \,\middle|\, n \in \mathbf{Z} \right\}. \tag{1.19}$$

If $-1 \in \Gamma$, we have $\begin{bmatrix} 1 & h \\ 0 & 1 \end{bmatrix} \in \rho\Gamma_s\rho^{-1}$. If $-1 \notin \Gamma$, however, there are two possibilities; $\rho\Gamma_s\rho^{-1}$ is generated by $\begin{bmatrix} 1 & h \\ 0 & 1 \end{bmatrix}$ or by $-\begin{bmatrix} 1 & h \\ 0 & 1 \end{bmatrix}$. We say that s is a **regular cusp** or an **irregular cusp** of Γ accordingly. This definition does not depend on the choice of ρ. (We can define these with $\rho \in SL_2(\mathbf{R})$. Since our group Γ is contained in $\Gamma(1)$, we can restrict ρ to $\Gamma(1)$. Then h is always a positive integer.)

1.11. We fix a congruence subgroup of $\Gamma(1)$ and let ν_2 resp. ν_3 denote the number of Γ-inequivalent elliptic points of order 2 resp. 3. Further let m be the number of orbits in $\Gamma \backslash (\mathbf{Q} \cup \{\infty\})$. Then the genus g of the compact Riemann surface $\Gamma \backslash (H \cup \mathbf{Q} \cup \{\infty\})$ is determined by

$$(1.20)\qquad 12(g-1) = \bigl[\Gamma(1) : \{\pm 1\}\Gamma\bigr] - 3\nu_2 - 4\nu_3 - 6m;$$

see [S71, Proposition 1.40].

When $-1 \notin \Gamma$, let u resp. u' denote the number of Γ-inequivalent regular resp. irregular cusps of Γ. Clearly $m = u + u'$.

Since $f\|_k(-1) = (-1)^k f$, we see that $\mathscr{M}_k(\Gamma) = \{0\}$ for odd k if $-1 \in \Gamma$. Therefore, whenever we consider $\mathscr{M}_k(\Gamma)$ for odd k, we assume that $-1 \notin \Gamma$. We now state the formula for $\dim \mathscr{M}_k(\Gamma)$ and $\dim \mathscr{M}_k(\Gamma) - \dim \mathscr{S}_k(\Gamma)$ for $0 < k \in \mathbf{Z}$:

$$dim\mathscr{M}_k(\Gamma) = \begin{cases} (k-1)(g-1) + km/2 + \varepsilon_k & (2 < k \in 2\mathbf{Z}), \\ g + m - 1 & (k = 2), \\ (k-1)(g-1) + uk/2 + u'(k-1)/2 + \varepsilon_k & (0 < k-1 \in 2\mathbf{Z}), \end{cases}$$

$$\varepsilon_k = \nu_1[k/4] + \nu_3[k/3],$$

$$\dim \mathscr{M}_k(\Gamma) - \dim \mathscr{S}_k(\Gamma) = \begin{cases} m & (2 < k \in 2\mathbf{Z}), \\ m-1 & (k = 2), \\ u & (0 < k-1 \in 2\mathbf{Z}), \\ u/2 & (k = 1). \end{cases}$$

The formula for $\dim \mathscr{M}_k(\Gamma)$ is included in [S71, Theorems 2.23 and 2.25]. That for $\dim \mathscr{M}_k(\Gamma) - \dim \mathscr{S}_k(\Gamma)$ can be derived easily from those theorems except in the case $k = 1$. To prove the case $k = 1$, let the symbols B, Q_j, Q'_j, and η be as in [S71, p. 47]. For $k = 1$ we have $[B] = 2^{-1}\mathrm{div}(\eta) + 2^{-1}\sum_{j=1}^{u} Q_j$ and

$$\bigl[B - \textstyle\sum_{j=1}^{u} Q_j - 2^{-1}\sum_{j=1}^{u'} Q'_j\bigr] = 2^{-1}\mathrm{div}(\eta) - 2^{-1}\sum_{j=1}^{u} Q_j = \mathrm{div}(\eta) - [B].$$

By the Riemann-Roch theorem we have

$$(*)\qquad \ell([B]) = \deg\bigl([B]\bigr) - g + 1 + \ell\bigl(\mathrm{div}(\eta) - [B]\bigr).$$

Now $\dim \mathscr{M}_k(\Gamma) = \ell([B])$ and $\dim \mathscr{S}_k(\Gamma) = \ell\bigl(\bigl[B - \sum_{j=1}^{u} Q_j - 2^{-1}\sum_{j=1}^{u'} Q'_j\bigr]\bigr)$ by [S71, (2.6.1), (2.6.2), and Lemma 2.21]. Since $\deg\bigl(\mathrm{div}(\eta)\bigr) = 2g-2$, from $(*)$ we obtain $\dim \mathscr{M}_k(\Gamma) = u/2 + \dim \mathscr{S}_k(\Gamma)$ as expected. Thus u is always even.

Finally we note that

$$(1.21)\qquad \dim \mathscr{M}_k(\Gamma) = u/2 \quad\text{and}\quad \mathscr{S}_k(\Gamma) = \{0\} \quad\text{if}\quad u > 2g-2,$$

as stated in [S71, (2.6.7), (2.6.8)].

Let us now discuss, as easy examples, regular and irregular cusps in the case $\Gamma = \Gamma_1(4)$. We see that $\Gamma_0(4) = \{\pm 1\}\Gamma_1(4)$, and so by [S71, Proposition 1.43] we have $\nu_2 = \nu_3 = 0$ and $m = 3$ for this group, and consequently $g = 0$. Since u is even and ≤ 3, we have $u = 2$ and $u' = 1$. Then we easily see that the above formula for $\dim \mathscr{M}_k(\Gamma)$ takes the form

$$(1.22)\qquad \dim \mathscr{M}_k\bigl(\Gamma_1(4)\bigr) = 1 + [k/2] \qquad (0 < k \in \mathbf{Z}).$$

Take the cusp $s = 1/2$. Then, for $\rho = \begin{bmatrix} 1 & 0 \\ -2 & 1 \end{bmatrix}$ we have

$$-\rho^{-1}\begin{bmatrix} 1 & 1 \\ 0 & 1 \end{bmatrix}\rho = \begin{bmatrix} 1 & -1 \\ 4 & -3 \end{bmatrix} \in \Gamma_1(4)_s.$$

This shows that $-\begin{bmatrix} 1 & 1 \\ 0 & 1 \end{bmatrix} \in \rho\Gamma_1(4)_s\rho^{-1}$, and $h = 1$ in the present case, as $0 < h \in \mathbf{Z}$. Thus we can conclude that $s = 1/2$ is an irregular cusp, and so the remaining two Γ-classes of cusps must be regular.

The group $\Gamma_1(4)$ is neither special nor typical in the sense explained in Theorem 1.13 below. We first prove

Lemma 1.12. *Let χ be a primitive Dirichlet character of conductor $2g$ with $0 < g \in \mathbf{Z}$. Then $\chi(1-g) = -1$.*

PROOF. Since g is even, we have $(1-g)^2 - 1 \in 2g\mathbf{Z}$, and so $\chi(1-g) = \pm 1$. Suppose $\chi(1-g) = 1$; let $x = 1 - yg$ with $y \in \mathbf{Z}$. Then $x - (1-g)^y \in 2g\mathbf{Z}$, and so $\chi(x) = 1$, which is a contradiction, as $2g$ is the conductor of χ. Thus $\chi(1-g) = -1$ as expected.

We will give a shorter, if nonelementary, proof of this lemma in the proof of Theorem 4.14 (ii).

Theorem 1.13. (i) *$\Gamma_1(N)$ has no irregular cusp if $N \neq 4$.*

(ii) *Let $\Gamma = \{\gamma \in \Gamma_0(N) \mid \chi(d_\gamma) = 1\}$ with a real character χ of $(\mathbf{Z}/N\mathbf{Z})^\times$ such that $\chi(-1) = -1$, and let t be the conductor of χ. Then Γ has an irregular cusp exactly in the following two cases:* (a) *t is even and N/t is an odd integer;* (b) *$8|t$ and N is $2t$ times an odd integer.*

PROOF. Suppose $\Gamma_1(N)$ has an irregular cusp. Then $-\rho^{-1}\begin{bmatrix} 1 & h \\ 0 & 1 \end{bmatrix}\rho \in \Gamma_1(N)$ with some $\rho \in \Gamma(1)$ and $0 < h \in \mathbf{Z}$. Put $\rho = \begin{bmatrix} a & b \\ c & d \end{bmatrix}$ and $\alpha = -\rho^{-1}\begin{bmatrix} 1 & h \\ 0 & 1 \end{bmatrix}\rho$. Then

$$\alpha = \begin{bmatrix} -1 - cdh & -d^2h \\ c^2h & cdh - 1 \end{bmatrix}. \tag{1.23}$$

This belongs to $\Gamma_1(N)$ only if $cdh + 2 \in N\mathbf{Z}$ and $cdh - 2 \in N\mathbf{Z}$, in which case $4 \in N\mathbf{Z}$. This is impossible if $N \neq 4$, as $\Gamma(1)$ and $\Gamma(2)$ contain -1. Thus we obtain (i).

To prove (ii), we first note that $-1 \notin \Gamma$ and $\Gamma_0(N) = \{\pm 1\}\Gamma$ for Γ defined there. Our setting means that $t|N$. Suppose Γ has an irregular cusp. Then we can find ρ and h such that α defined as above belongs to Γ, and (1.23) shows that $t|c^2h$. Suppose t is odd; then t must be squarefree, as χ is real. Thus $t|ch$, and $\chi(cdh - 1) = \chi(-1) = -1$, which is impossible, as $\alpha \in \Gamma$. Therefore t is even, and so $t = 4r$ or $t = 8r$ with an odd integer r. If $t|ch$, we are again led to a contradiction, and so $t \nmid ch$, though $t | c^2h$. Since $2r$ is squarefree and $2r | c^2h$, we have $2r | ch$. Suppose c is odd. Then $4 | h$ or $8 | h$ according as $t = 4r$ or $t = 8r$. Since $r | ch$, we have $t | ch$, a contradiction. Thus we can put $c = 2x$ with $x \in \mathbf{Z}$. We now proceed according to the nature of t and the parity of x.

(1) Suppose $t = 4r$; suppose also that x is even. Then $4 | c$ and $4r | ch$, as $r | ch$. This is a contradiction. Thus x is odd if $t = 4r$. Since $r | x^2h$, we have $r | xh$. If h is even, then $4 | ch$, and so $4r | ch$, a contradiction. Thus h is odd. Since $4r | N$, $N | 4x^2h$, and xh is odd, we see that N is $4r$ times an odd integer. This falls into Case (a) of (ii).

(2) Next suppose $t = 8r$; then $2r | x^2h$. If x is odd, then $2 | h$, but $4 \nmid h$, as $8r \nmid ch$. Thus $h = 2g$ with an odd integer g. This time $8r | N$ and $N | 8x^2g$. Since xg is odd, we see that N is $8r$ times an odd integer. This falls into Case (a) of (ii).

(3) It remains to consider the case in which $t = 8r$ and x is even. Put $x = 2y$. Then $r | 2y^2h$ and $2r \nmid yh$. Since $r | yh$, we see that yh is odd. Now $8r | N$ and $N | 16y^2h$. Therefore N is either $8r$ times an odd integer or $16r$ times an odd integer. Consequently (N, t) falls into Case (a) or (b). This completes the proof of the "only if"-part of (ii).

Conversely, suppose (N, t) is of type (a) of (ii). Then we can put $N = (2e+1)t$ with $e \in \mathbf{Z}$. Take $\rho = \begin{bmatrix} 1 & 0 \\ 2 & 1 \end{bmatrix}$ (that is, $c = 2$ and $d = 1$) and $h = (2e+1)t/4$. Then $c^2h = N$, $cdh = (2e+1)t/2$, and $\chi(cdh - 1) = \chi(t/2 - 1) = 1$ by Lemma 1.12, as $\chi(-1) = -1$. Thus $\alpha \in \Gamma$, which means that $1/2$ is an irregular cusp.

Next supose $t = 8r$ and $N = 2t(2f+1)$ with $f \in \mathbf{Z}$. Take $\rho = \begin{bmatrix} 1 & 0 \\ 4 & 1 \end{bmatrix}$ (that is, $c = 4$ and $d = 1$) and $h = r(2f+1)$. Then $c^2h = N$, $cdh = 4r(2f+1)$, and $\chi(cdh - 1) = \chi(4r - 1) = 1$ by Lemma 1.12. Thus $\alpha \in \Gamma$, and so $1/4$ is an irregular cusp. This completes the proof of our theorem.

Take $N = t = 8$ in (ii), for example. Then $\Gamma_0(8) = \{\pm 1\}\Gamma$. In this case $m = 4$ with m as in §1.11. Since ∞ is aregular cusp, we obtain $u = u' = 2$ in this case.

2. Elementary Fourier analysis

2.1. Let us first recall some basic facts on the gamma function $\Gamma(s)$. It is a meromorphic function on $\mathbf{C}$ with the following properties:

(2.1a) $$\Gamma(s) = \int_0^\infty e^{-t}t^{s-1}dt \quad \text{if} \quad \mathrm{Re}(s) > 0,$$

(2.1b) $$\Gamma(s+1) = s\Gamma(s),$$

(2.1c) $$\Gamma(s)^{-1} \quad \textit{is an entire function,}$$

(2.1d) *The set of poles of Γ consists of 0 and all negative integers, and each pole is of order 1,*

(2.1e) $$\Gamma(n) = (n-1)! \quad \textit{if} \quad 0 < n \in \mathbf{Z},$$

(2.1f) $$\Gamma(1/2) = \sqrt{\pi},$$

(2.1g) $$\Gamma(m + 1/2) \in \sqrt{\pi}\,\mathbf{Q}^\times.$$

The last fact follows from (2.1b) and (2.1f). Substituting at for t in (2.1a), we obtain the following formula at least when $0 < a \in \mathbf{R}$:

(2.1h) $$\Gamma(s)a^{-s} = \int_0^\infty e^{-at}t^{s-1}dt \quad \textit{if} \quad a \in \mathbf{C},\ \mathrm{Re}(a) > 0,\ \textit{and}\ \mathrm{Re}(s) > 0.$$

Here $a^{-s} = \exp(-s\log a)$ with the standard branch of $\log a$ for $\mathrm{Re}(a) > 0$. Indeed, the integral is meaningful for $\mathrm{Re}(a) > 0$, and defines a holomorphic function of a. (See Theorem A1.4 of the Appendix.) Since it coincides with $\Gamma(s)a^{-s}$ for $0 < a \in \mathbf{R}$, we obtain (2.1h) as stated.

In this section, or rather in this whole book, a function is a $\mathbf{C}$-valued function, unless stated otherwise. For $f \in L^1(\mathbf{R}^n)$ we define its **Fourier transform** $\hat{f}$ by

(2.2a) $$\hat{f}(x) = \int_{\mathbf{R}^n} f(y)\mathbf{e}(-{}^t xy)dy \qquad (x \in \mathbf{R}^n),$$

where we consider x and y column vectors, so that ${}^t xy = \sum_{\nu=1}^n x_\nu y_\nu$. Then $\hat{f}$ is a continuous function; moreover if $\hat{f} \in L^1(\mathbf{R}^n)$, then

(2.2b) $$f(x) = \int_{\mathbf{R}^n} \hat{f}(y)\mathbf{e}({}^t xy)dy$$

holds almost everywhere. This is called the **Fourier inversion formula.** If $f \in L^1(\mathbf{R}^n) \cap L^2(\mathbf{R}^n)$, then $\hat{f} \in L^2(\mathbf{R}^n)$ and f has the same L^2-norm as $\hat{f}$, that is,

(2.2c) $$\int_{\mathbf{R}^n} |\hat{f}(y)|^2 dy = \int_{\mathbf{R}^n} |f(x)|^2 dx.$$

This is called the **Plancherel formula.** For these and other facts on Fourier analysis stated in this section but not proved, the reader is referred to any textbook on real analysis or Fourier analysis, and also to Section A2 of the Appendix.

It is well known that the function $\exp(-\pi x^2)$ is its own Fourier transform, that is,

(2.3) $$\int_{\mathbf{R}} \exp(-\pi x^2)\mathbf{e}(-xt)dx = \exp(-\pi t^2).$$

We give here a short proof as follows. We recall the formula $\int_{-\infty}^{\infty} \exp(-ax^2)dx = \sqrt{\pi/a}$, or equivalently, $\int_{-\infty}^{\infty} \exp(-\pi x^2)dx = 1$, which can be found any elementary calculus book. We now prove

$$\int_{\mathbf{R}} \exp\big(-\pi(x+z)^2\big)dx = 1 \qquad \text{for every } z \in \mathbf{C}. \tag{2.4}$$

Indeed, for $z = u + it$ with real u and t we have

$$\exp\big(-\pi(x+u+it)^2\big) = \exp\big(-\pi(x+u)^2\big)\exp\big(-2\pi(x+u)it\big)\exp(\pi t^2).$$

This shows that the integral of (2.4) is uniformly convergent for z in a compact subset of $\mathbf{C}$. Therefore the left-hand side of (2.3) is holomorphic in z. (See Theorem A1.4 of the Appendix.) Also, the value of the integral is 1 if $z \in \mathbf{R}$. This proves (2.4), as a holomorphic function on $\mathbf{C}$ is determined by its values on $\mathbf{R}$. Taking $z = it$, we obtain (2.3) from (2.4).

2.2. Let L be a lattice in $\mathbf{R}^n$, that is, a discrete subgroup of $\mathbf{R}^n$ isomorphic to $\mathbf{Z}^n$, and let

$$\tilde{L} = \big\{x \in \mathbf{R}^n \;\big|\; {}^t xy \in \mathbf{Z} \;\; \textit{for every} \;\; y \in L\big\}. \tag{2.5}$$

We easily see that $\mathrm{vol}(\mathbf{R}^n/\tilde{L}) = \mathrm{vol}(\mathbf{R}^n/L)^{-1}$. Given $g \in L^2(F)$, $F = \mathbf{R}^n/L$, we define the **Fourier coefficient** c_ξ of g for each $\xi \in \tilde{L}$ by

$$c_\xi = \mathrm{vol}(F)^{-1}\int_F g(x)\mathbf{e}(-{}^t\xi x)dx. \tag{2.6}$$

This is meaningful, as $\mathbf{e}({}^t\xi x)$ defines an element of $L^2(F)$ and the integral is an inner product of that element and g. Moreover,

$$\mathrm{vol}(F)\sum_{\xi\in\tilde{L}} |c_\xi|^2 = \int_F |g(x)|^2 dx, \tag{2.7}$$

which is similar to (2.2c). Now we have

$$g(x) = \sum_{\xi\in\tilde{L}} c_\xi \mathbf{e}({}^t\xi x) \quad \textit{if } g \textit{ is continuous on } F \textit{ and} \sum_{\xi\in\tilde{L}} |c_\xi| < \infty. \tag{2.8}$$

In fact, put $h(x) = g(x) - \sum_{\xi\in\tilde{L}} c_\xi \mathbf{e}({}^t\xi x)$. Observe that h is continuous on F if g is continuous and $\sum_{\xi\in\tilde{L}} |c_\xi| < \infty$. Clearly all the Fourier coefficients of h are 0. Therefore (2.7) applied to h shows that $h = 0$. This proves (2.8). From this we can derive

Theorem 2.3 (Poisson Summation Formula). *Let f be a continuous function belonging to $L^1(\mathbf{R}^n)$. Then*

$$\mathrm{vol}(\mathbf{R}^n/L)\sum_{\ell\in L} f(x+\ell) = \sum_{m\in\tilde{L}} \hat{f}(m)\mathbf{e}({}^t mx), \tag{2.9}$$

provided both sides converge absolutely and uniformly. (As for the right-hand side, this means the convergence of $\sum_{m\in\tilde{L}} |\hat{f}(m)|$.) Especially,

$$\text{(2.10)} \qquad \mathrm{vol}(\mathbf{R}^n/L)\sum_{\ell\in L} f(\ell) = \sum_{m\in\tilde{L}} \hat{f}(m).$$

PROOF. Put $g(x) = \sum_{\ell\in L} f(x+\ell)$. This is continuous on F. Then for $m \in \tilde{L}$ we have

$$\int_F g(x)\mathbf{e}(-{}^t mx)dx = \int_F \sum_{\ell\in L} f(x+\ell)\mathbf{e}(-{}^t m(x+\ell))dx$$
$$= \int_{\mathbf{R}^n} f(x)\mathbf{e}(-{}^t mx)dx = \hat{f}(m).$$

Hence (2.8) gives our formulas.

Remark. The absolute convergence of the right-hand side of (2.9) is guaranteed if the left-hand side defines a C^r function with $r > n/2$; see Theorem A2.2 of the Appendix.

2.4. For fixed $s \in \mathbf{C}$ and $z \in H$ define f by $f(t) = t^{s-1}\mathbf{e}(tz)$ for $t > 0$ and $f(t) = 0$ for $t \le 0$. This is integrable if $\mathrm{Re}(s) > 0$ and continuous if $\mathrm{Re}(s) > 1$. By (2.1h) we find that $\hat{f}(u) = \Gamma(s)[-2\pi i(z-u)]^{-s}$ if $\mathrm{Re}(s) > 0$. We see that $f \in L^2(\mathbf{R})$ if $\mathrm{Re}(s) > 1/2$, and so the general principle of the Fourier transform says that $\hat{f} \in L^2(\mathbf{R})$ if $\mathrm{Re}(s) > 1/2$, which means that $[-2\pi i(z-u)]^{-2s}$ is integrable for $\mathrm{Re}(s) > 1/2$. Consequently, $\hat{f} \in L^1(\mathbf{R})$ if $\mathrm{Re}(s) > 1$. Applying (2.2b) to this case, we obtain

$$\text{(2.11)} \qquad \Gamma(s)\int_{-\infty}^{\infty} [-2\pi i(z-u)]^{-s}\mathbf{e}(tu)du = \begin{cases} t^{s-1}\mathbf{e}(tz) & (t > 0), \\ 0 & (t \le 0) \end{cases}$$

for $\mathrm{Re}(s) > 1$, which can be written also

$$\text{(2.12)} \qquad \Gamma(s)\int_{-\infty}^{\infty} (y+ix)^{-s}e^{itx}dx = \begin{cases} 2\pi \cdot t^{s-1}e^{-ty} & (t > 0), \\ 0 & (t \le 0) \end{cases}$$

for $\mathrm{Re}(s) > 1$, $y > 0$. Now (2.10) applied to this case produces

$$\text{(2.13)} \quad \sum_{n=1}^{\infty} n^{s-1}\mathbf{e}(nz) = \Gamma(s)\sum_{m\in\mathbf{Z}} [-2\pi i(z-m)]^{-s} \qquad (\mathrm{Re}(s) > 1,\ z \in H).$$

In particular, taking $s = k \in \mathbf{Z}$, we obtain

$$\text{(2.14)} \quad (k-1)!\sum_{m\in\mathbf{Z}} (z+m)^{-k} = (-2\pi i)^k \sum_{n=1}^{\infty} n^{k-1}\mathbf{e}(nz) \quad (1 < k \in \mathbf{Z},\ z \in H).$$

Either of these two equalities is called the **Lipschitz formula.**

We insert here the proof of Lemma 1.6 (ii). By (2.13) we have

$$\sum_{n=1}^{\infty} n^{\alpha}e^{-2\pi ny} = \Gamma(\alpha+1)\sum_{m\in\mathbf{Z}} [2\pi y + 2\pi im]^{-\alpha-1} \qquad (0 < y \in \mathbf{R},\ 0 < \alpha \in \mathbf{R}).$$

The last sum minus the term for $m = 0$ is uniformly convergent if $\alpha > 0$ and y is small, and so the sum is $O(y^{-\alpha-1})$ as $y \to 0$, which proves the desired fact.

Putting

$$\mathbf{q} = \mathbf{e}(z) \qquad (z \in \mathbf{C}) \tag{2.15}$$

for simplicity, we can extend (2.14) from H to $\mathbf{C}$ as follows:

$$(k-1)! \sum_{m \in \mathbf{Z}} (z+m)^{-k} = (2\pi i)^k \cdot \frac{\mathbf{q}P_k(\mathbf{q})}{(\mathbf{q}-1)^k} \qquad (1 < k \in \mathbf{Z},\ z \in \mathbf{C},\ \notin \mathbf{Z}) \tag{2.16}$$

with a polynomial P_k of degree $k-2$ with coefficients in $\mathbf{Z}$, which satifies a recurrence formula

$$P_{k+1} = (kx - x + 1)P_k - (x^2 - x)P_k', \qquad P_2(x) = 1. \tag{2.17}$$

Indeed, we have $\sum_{n=1}^{\infty} n\mathbf{q}^n = \mathbf{q}/(1-\mathbf{q})^2$, which combined with (2.14) gives (2.16) for $z \in H$ in the case $k = 2$. Applying $d/d\mathbf{q}$ successively to that case, we obtain (2.16) for $z \in H$ with P_k satisfying (2.17). Since both sides of (2.16) are holomorphic in $z \in \mathbf{C}, \notin \mathbf{Z}$, we obtain (2.16) for such z and $k > 1$.

In fact, we have a formula even for $k = 1$:

$$\pi i \cdot \frac{\mathbf{e}(z)+1}{\mathbf{e}(z)-1} = z^{-1} + \sum_{n=1}^{\infty} \{(z+n)^{-1} + (z-n)^{-1}\} \quad (z \in \mathbf{C}, \notin \mathbf{Z}). \tag{2.18}$$

This is classical and can be found in most textbooks on complex analysis. Once (2.18) is established, we obtain (2.16) by applying d^{k-1}/dz^{k-1} to (2.18). Conversely, (2.18) can be derived from (2.16) as follows. Since equality (2.16) for $k = 2$ is "the derivative" of (2.18), we see that (2.18) is true up to addition of a constant. The right-hand side of (2.18) equals

$$z^{-1} + \sum_{n=1}^{\infty} \frac{2z}{z^2 - n^2} = z^{-1} - \sum_{n=1}^{\infty} \frac{2n^{-2}z}{1 - n^{-2}z^2} = z^{-1} - 2\sum_{m=0}^{\infty} z^{2m+1} \sum_{n=1}^{\infty} n^{-2m-2}.$$

The left-hand side of (2.18) is $\pi i + 2\pi i \big(\mathbf{e}(z) - 1\big)^{-1}$, which equals

$$\pi i + z^{-1}(1 + \pi i z + \cdots)^{-1}.$$

Thus the Laurent expansions of both sides of (2.18) have 0 as their constant terms. This proves (2.18).

Here are the explicit forms of P_k for $3 \le k \le 5$:

$$P_3(x) = x + 1, \quad P_4(x) = x^2 + 4x + 1, \quad P_5(x) = x^3 + 11x^2 + 11x + 1.$$

We will prove in §4.4 the formula

$$x^{n-2}P_n(x^{-1}) = P_n(x). \tag{2.19}$$

We will also prove in §§4.4 and 9.1 two recurrence formulas:

$$(2.20)\quad P_{n+1}(x)=\sum_{k=0}^{n-1}\binom{n}{k}(x-1)^{n-1-k}P_{k+1}(x)\qquad (n>0,\ P_1(x)=1),$$

$$(2.21)\quad P_{n+5}(x)-(x-1)^2P_{n+3}(x)=12x\sum_{a=0}^{n}\binom{n}{a}P_{a+3}(x)P_{n-a+2}(x)\quad (n\geq 0).$$

2.5. We now consider a theta function of the simplest type:

$$(2.22)\qquad \theta_N^r(z,\,a)=\sum_{n\in a+N\mathbf{Z}} n^r\mathbf{e}(n^2z/2)\qquad (z\in H).$$

Here $r=0$ or 1, $0<N\in\mathbf{Z}$, $a\in\mathbf{Z}$, and $a+N\mathbf{Z}=\{n\in\mathbf{Z}\,|\,n-a\in N\mathbf{Z}\}$. The convergence of the last infinite series can easily be seen, as $|\mathbf{e}(n^2z/2)|=\exp(-\pi n^2y)$ with $y=\mathrm{Im}(z)>0$. We have a tranformation formula

$$(2.23)\quad \theta_N^r(-(N^2z)^{-1},\,a)=(-iN)^r(-iz)^{(2r+1)/2}\sum_{b=1}^{N}\mathbf{e}(ab/N)\theta_N^r(z,b)$$
$$=(-iN)^r(-iz)^{(2r+1)/2}\sum_{n\in\mathbf{Z}}\mathbf{e}(na/N)n^r\mathbf{e}(n^2z/2),$$

where $(-iz)^s$ denotes the branch that becomes y^s when $z=iy$, $y>0$. To prove (2.23), we first note

$$(2.24)\quad \int_{\mathbf{R}}\mathbf{e}(u^2z/2)\mathbf{e}(-tu)du=(-iz)^{-1/2}\mathbf{e}(-t^2z^{-1}/2)\qquad (z\in H,\ t\in\mathbf{R}).$$

Indeed, this equality is true if $z=iy$, $y>0$, by virtue of (2.3). Since both sides are holomorphic in z, we obtain (2.24). Applying $\partial/\partial t$ to (2.24) and combining the result with the original (2.24), we obtain

$$(2.25)\quad \int_{\mathbf{R}}u^r\mathbf{e}(-u^2z^{-1}/2)\mathbf{e}(-tu)du=i^{-r}(-iz)^{(2r+1)/2}t^r\mathbf{e}(t^2z/2)\quad (r=0,\,1).$$

Differentiation can be justified by Theorem A1.3 (2) of the Appendix. Take $f(x)=x^r\mathbf{e}(-x^2z^{-1}/2)$ and $L=N\mathbf{Z}$ in (2.9). Then the left-hand side of (2.9) for $x=a$ is

$$N\sum_{n\in\mathbf{Z}}(a+Nn)^r\mathbf{e}(-(a+Nn)^2z^{-1}/2)=N\theta_N^r(-z^{-1},\,a).$$

Since $\tilde{L}=N^{-1}\mathbf{Z}$, the right-hand side of (2.9) for $x=a$ is $\sum_{h\in\mathbf{Z}}\mathbf{e}(ah/N)\hat{f}(h/N)$. From (2.25) we obtain $\hat{f}(t)=i^{-r}(-iz)^{(2r+1)/2}t^r\mathbf{e}(t^2z/2)$. Therefore, substituting N^2z for z, we see that (2.9) in the present setting can be written in the form (2.23)

In particular, in the case $N=1$ put

$$(2.26)\qquad \theta(z)=\theta_1^0(2z,\,0)=\sum_{n\in\mathbf{Z}}\mathbf{e}(n^2z)\qquad (z\in H).$$

Then from (2.23) we obtain

$$\theta(-(4z)^{-1}) = (-2iz)^{1/2}\theta(z) \qquad (z \in H). \tag{2.27}$$

Another good example is

$$\eta(z) = \sum_{n=1}^{\infty} \left(\frac{3}{n}\right)\mathbf{e}(n^2z/24) \qquad (z \in H). \tag{2.28}$$

We have $2\eta(12z) = \sum_{a=1}^{12} \psi(a)\theta_{12}^0(z, a)$, where $\psi(a) = \left(\frac{3}{a}\right)$. Taking $r = 0$ and $N = 12$ in (2.23), we can verify, after a few lines of easy calculation, the first of the following two equalities:

$$\eta(-z^{-1}) = (-iz)^{1/2}\eta(z), \qquad \eta(z+1) = \mathbf{e}(1/24)\eta(z). \tag{2.29}$$

The latter equality follows from the fact that $n^2 - 1 \in 24\mathbf{Z}$ if n is prime to 6. Since $\Gamma(1)$ is generated by $\begin{bmatrix} 0 & -1 \\ 1 & 0 \end{bmatrix}$ and $\begin{bmatrix} 1 & 1 \\ 0 & 1 \end{bmatrix}$, from (2.29) we see that $\eta(\gamma z) = \zeta_\gamma j_\gamma(z)^{1/2}\eta(z)$ for every $\gamma \in \Gamma(1)$ with a 24th root of unity ζ_γ. Consequently η^{24} belongs to $\mathscr{M}_{12}(\Gamma(1))$ and even to $\mathscr{S}_{12}(\Gamma(1))$, as can easily be seen. Since that space of cusp forms is generated by Δ, we can conclude that $\eta^{24} = \Delta$. This η is traditionally called **Dedekind's eta function**. In general, it can be shown that $\theta_N^r(qz, a)$ for $0 < q \in \mathbf{Q}$ is a modular form of weight $r + 1/2$; see [S73].

3. The functional equation of a Dirichlet series

3.1. Given a function $f(z) = \sum_{n=0}^{\infty} a_n\mathbf{e}(nz/N)$ on H, we are going to associate a Dirichlet series $\sum_{n=1}^{\infty} a_n n^{-s}$. To state the results in the form suitable for our later applications, we take f to be a holomorphic function on H given in the form

$$f(z) = \sum_{0 \le \xi \in \mathbf{Q}} a_\xi \mathbf{e}(\xi z), \quad a_\xi = 0 \quad \text{for} \quad \xi \notin h^{-1}\mathbf{Z} \tag{3.1}$$

with $a_\xi \in \mathbf{C}$ and some $h \in \mathbf{Z}, > 0$. This f may or may not be a modular form. Fixing a positive integer N and a real number k, we put

$$f^{\#}(z) = N^{-k/2}(-iz)^{-k}f\bigl(-(Nz)^{-1}\bigr), \tag{3.2}$$

where the branch of $(-iz)^{-k}$ is taken so that it is real and positive if $z \in H \cap i\mathbf{R}$, and assume that $f^{\#}(z)$ can be written in the form $f^{\#}(z) = \sum_{0 \le \xi \in \mathbf{Q}} b_\xi \mathbf{e}(\xi z)$, $b_\xi = 0$ for $\xi \notin h^{-1}\mathbf{Z}$. We also assume that $|a_\xi| = O(\xi^\alpha)$ and $|b_\xi| = O(\xi^\alpha)$ as $\xi \to \infty$ with some $\alpha \in \mathbf{R}$, and put

$$D(s, f) = \sum_{\xi > 0} a_\xi \xi^{-s}, \qquad R(s, f) = (2\pi)^{-s}\Gamma(s)D(s, f). \tag{3.3}$$

Clearly the sum over $\xi > 0$ converges for $\mathrm{Re}(s) > \alpha + 1$, and so $D(s, f)$ is meaningful as a holomorphic function on that half plane; the same is true for $D(s, f^{\#})$.

We can take any element of $\mathscr{M}_k$ to be f, as the estimate of a_ξ and b_ξ holds by virtue of Lemma 1.8.

Theorem 3.2. *In the above setting $R(s, f)$ and $R(s, f^\#)$ can be continued as meromorphic functions to the whole s-plane with the following properties:*

$$N^{s/2}R(s, f) = -\frac{a_0}{s} + \frac{b_0}{s-k} + \text{an entire function}, \tag{3.4a}$$

$$R(k-s, f) = N^{s-k/2}R(s, f^\#). \tag{3.4b}$$

In particular, $R(s, f)$ is an entire function if $a_0 = b_0 = 0$.

PROOF. We easily see that $(f^\#)^\# = f$. Put $F(y) = f(iy) - a_0$ and $G(y) = f^\#(iy) - b_0$. Then termwise integration shows, in view of (2.1h), that

$$\int_0^\infty F(y)y^{s-1}dy = \sum_{\xi>0} a_\xi \int_0^\infty e^{-2\pi\xi y}y^{s-1}dy = \Gamma(s)(2\pi)^{-s}\sum_{\xi>0} a_\xi\xi^{-s} = R(s, f).$$

This is justified for $\mathrm{Re}(s) > \alpha + 1$, because of the convergence of $\sum_{n=1}^\infty |a_\xi\xi^{-s}|$. (See (A1.2) of the Appendix.) Put $p = N^{-1/2}$. We have

$$R(k-s, f) = \int_0^p F(y)y^{k-s-1}dy + \int_p^\infty F(y)y^{k-s-1}dy$$

at least for $\mathrm{Re}(k-s) > \alpha + 1$. Substituting $(Ny)^{-1}$ for y, we obtain

$$\int_0^p F(y)y^{k-s-1}dy = N^{s-k}\int_p^\infty F\big((Ny)^{-1}\big)y^{s-k-1}dy. \tag{$*$}$$

Now $F\big((Ny)^{-1}\big) = f\big(i(Ny)^{-1}\big) - a_0 = N^{k/2}y^k\big[G(y) + b_0\big] - a_0$. Thus $(*)$ equals

$$N^{s-k/2}\int_p^\infty G(y)y^{s-1}dy + N^{s-k/2}b_0\int_p^\infty y^{s-1}dy - N^{s-k}a_0\int_p^\infty y^{s-k-1}dy.$$

The last two terms are $-b_0N^{(s-k)/2}/s$ and $a_0N^{(s-k)/2}/(s-k)$, provided $\mathrm{Re}(s) < 0$. Therefore

$$R(k-s, f) = \int_p^\infty F(y)y^{k-s-1}dy + N^{s-k/2}\int_p^\infty G(y)y^{s-1}dy + \frac{a_0N^{(s-k)/2}}{s-k} - \frac{b_0N^{(s-k)/2}}{s}$$

for $\mathrm{Re}(s) < \mathrm{Min}\{0, k-\alpha-1\}$. In view of Lemma 1.6 (i) we see that the two integrals over $[p, \infty)$ are convergent for every $s \in \mathbf{C}$ and define entire functions of s. Consequently $N^{(k-s)/2}R(k-s, f)$ can be continued to an entire function plus $a_0/(s-k) - b_0/s$. Take $f^\#$ in place of f and substitute $k-s$ for s. Thus, exchanging (F, a_0) for (G, b_0), we have

$$R(s, f^{\#}) = \int_p^\infty G(y)y^{s-1}dy + N^{k/2-s}\int_p^\infty F(y)y^{k-s-1}dy$$
$$-\frac{b_0 N^{-s/2}}{s} + \frac{a_0 N^{-s/2}}{s-k},$$

which equals $N^{k/2-s}R(k-s, f)$. This completes the proof.

3.3. We apply the above theorem to the series

$$M_{a,N}^{\nu}(s) = \sum_{0\neq n\in a+N\mathbf{Z}} n^{-\nu}|n|^{-2s}. \tag{3.5}$$

Here $\nu \in \mathbf{Z}$, $0 < N \in \mathbf{Z}$, and $a \in \mathbf{Z}$. Clearly this is convergent for $\mathrm{Re}(2s+\nu) > 1$. For every $c \in \mathbf{Z}$ we have

$$M_{a,N}^{\nu+2c}(s) = M_{a,N}^{\nu}(s+c). \tag{3.6}$$

We cannot extend this to the case $c \in 2^{-1}\mathbf{Z}$, as $|n|^{2c}$ may be different from n^{2c}. Thus we can reduce the problem about the analytic nature of the series to the case $\nu = -1$ or $\nu = 0$, according as ν is odd or even.

Theorem 3.4. *The function $(N/\pi)^s\Gamma(s)M_{a,N}^{\nu}(s)$ can be continued to the whole s-plane as a meromorphic function, which is entire if $\nu = -1$. If $\nu = 0$, it is an entire function plus $N^{-1/2}(s-2^{-1})^{-1} - \delta(a/N)s^{-1}$, where $\delta(x) = 1$ if $x \in \mathbf{Z}$, and $\delta(x) = 0$ otherwise.*

PROOF. Take $\theta_N^r(z, a)$ of (2.22) and N^2 as f and N of Theorem 3.2, and let $k = r + 1/2$. Then by (2.23) we have

$$f^{\#}(z) = i^{-r}N^{-1/2}\sum_{b=1}^{N}\mathbf{e}(ab/N)\theta_N^r(z, b).$$

Thus $a_0 = b_0 = 0$ if $r = 1$. For $r = 0$ we have $a_0 = \delta(a/N)$ and $b_0 = N^{-1/2}$. Also, $D(s, f) = 2^s M_{a,N}^{-r}(s)$ and $N^s R(s, f) = (N/\pi)^s\Gamma(s)M_{a,N}^{-r}(s)$. Therefore we obtain our theorem with $\nu = -r$ from Theorem 3.2. Then the first statement of our theorem follows from (3.6).

Taking $r = 0$ or 1, from (3.4b) we obtain

$$\pi^{s-k}\Gamma(k-s)M_{a,N}^{-r}(k-s) = R(k-s, f) = N^{2s-k}R(s, f^{\#}) \tag{3.7}$$
$$= N^{2s-r-1}\pi^{-s}\Gamma(s)i^{-r}\sum_{b=1}^{N}\mathbf{e}(ab/N)M_{b,N}^{-r}(s),$$

where $k = r + 1/2$. The sum $\sum_{b=1}^{N}$ can be written $\sum_{b=1}^{N-1}$ if $r = 1$, as $M_{0,N}^{-1} = 0$. Putting $s = 1$, we obtain $M_{a,N}^{-1}(1/2) = (\pi i)^{-1}\sum_{b=1}^{N-1}\mathbf{e}(ab/N)M_{b,N}^{-1}(1)$. Since $M_{a,N}^{1}(s) = M_{a,N}^{-1}(s+1)$, this proves (3.8) below.

Theorem 3.5. *Define* $M_{a,N}^{\nu}$ *by (3.5) with integers* a *and* N *such that* $0 < a < N$, *and put* $\zeta = \mathbf{e}(1/N)$. *Then*

$$M_{a,N}^{1}(-1/2) = \frac{1}{\pi i}\sum_{b=1}^{N-1}\zeta^{ab}M_{b,N}^{1}(0), \tag{3.8}$$

$$(\nu-1)!N^{\nu}M_{a,N}^{\nu}(0) = \begin{cases} \pi i(\zeta^a+1)(\zeta^a-1)^{-1} & (\nu=1), \\ (2\pi i)^{\nu}\zeta^a P_{\nu}(\zeta^a)(\zeta^a-1)^{-\nu} & (1<\nu\in\mathbf{Z}), \end{cases} \tag{3.9}$$

where P_{ν} *is the polynomial of (2.16) and (2.17).*

PROOF. To prove (3.9), we first consider the case $\nu = 1$. Taking z to be a/N with $0 < a < N$ in (2.18) we obtain

$$(\pi i/N)(\zeta^a+1)/(\zeta^a-1) = a^{-1} + \sum_{n=1}^{\infty}\{(a+nN)^{-1}+(a-nN)^{-1}\}. \tag{3.10}$$

To show that this equals $M_{a,N}^{1}(0)$, we observe that

$$M_{a,N}^{1}(s/2) = a^{-1-s} + \sum_{n=1}^{\infty}\{(a+nN)^{-1}|a+nN|^{-s}+(a-nN)^{-1}|a-nN|^{-s}\}$$

for $\mathrm{Re}(s) > 0$. The right-hand side with $s = 0$ becomes the right-hand side of (3.10), and so we obtain the desired formula for $M_{a,N}^{1}(0)$, provided this barbaric argument can be justified. To that end, we multiply the last equality by $-(1+s)^{-1}$. Thus

$$-\frac{M_{a,N}^{1}(s/2)}{1+s} = -\frac{a^{-1-s}}{1+s} + \sum_{n=1}^{\infty}\frac{1}{1+s}\{(nN-a)^{-1-s}-(nN+a)^{-1-s}\}.$$

The n-th term of the last infinite sum can be written $\int_{nN-a}^{nN+a}x^{-2-s}dx$. If $n > 1$, then $nN - a > 1$, and so the integral for $0 \le s \in \mathbf{R}$ can be majorized by $\int_{nN-a}^{nN+a}x^{-2}dx = (nN-a)^{-1}-(nN+a)^{-1} = 2a(n^2N^2-a^2)^{-1}$. Therefore the last infinite sum is uniformly convergent for $0 \le s \in \mathbf{R}$. This proves (3.9) for $\nu = 1$.

The case $\nu > 1$ is simpler. Substituting a/N for z in (2.16), we obtain (3.9) for $\nu > 1$.

3.6. We defined $M_{a,N}^{\nu}$ by (3.5) and formulated the results about its special values as in the above theorem, simply because they appear as constant terms of the Eisenstein series that we will discuss in Section 9. To state the formulas in a style closer to the traditional one, for $r = 0$ or 1 and $a \in \mathbf{Z}$, $0 \le a < N$, put

$$D_{a,N}^{r}(s) = \sum_{0\neq n\in a+N\mathbf{Z}} n^r|n|^{-r-s}. \tag{3.11}$$

It should be remembered that the sum is extended over both positive and negative n. We see that $D_{a,N}^{1} = 0$ if $a = 0$. For this reason we assume $0 < a < N$ if $r = 1$. Clearly

$$D^r_{a,N}(2s+2m-r) = M^{-r}_{a,N}(s+m) = M^{2m-r}_{a,N}(s), \tag{3.12}$$

and so $D^r_{a,N}(s) = M^{-r}_{a,N}\big((s+r)/2\big)$. Therefore from Theorem 3.4 we obtain the following theorem except (3.13a, b).

Theorem 3.7. *The product $(N/\pi)^{s/2}\Gamma\big((s+r)/2\big)D^r_{a,N}(s)$ can be continued to a meromorphic function on the whole s-plane, which is an entire function if $r=1$. If $r=0$, it is $2N^{-1/2}/(s-1) - 2\delta(a/N)/s$ plus an entire function. Moreover,*

$$D^r_{a,N}(r-2m) = 0 \quad \text{if} \quad 0 < m \in \mathbf{Z}, \tag{3.13a}$$

$$D^0_{a,N}(0) = -\delta(a/N). \tag{3.13b}$$

Formula (3.13a) follows from the fact that $\Gamma(s)$ has poles at $s=-m$ with $0 \le m \in \mathbf{Z}$; similarly (3.13b) follows from the analysis of $\Gamma(s/2)D^0_{a,N}(s)$ at $s=0$.

If $N=1$ and $r=a=0$, then $2^{-1}D^0_{0,1}(s)$ coincides with the **Riemann zeta function** $\zeta(s) = \sum_{n=1}^{\infty} n^{-s}$, and we obtain a few well known facts. Namely, put $\xi(s) = \pi^{-s/2}\Gamma(s)\zeta(s)$. Then

$$\xi(1-s) = \xi(s) = \frac{1}{s-1} - \frac{1}{s} + \text{an entire function.} \tag{3.14}$$

Since $D^r_{a,N}(2m-r) = M^{2m-r}_{a,N}(0)$, the value $D^r_{a,N}(2m-r)$ for $0 < m \in \mathbf{Z}$ and $r=0$ or 1 can be given explicitly by (3.9). We will give the formulas in Theorem 4.7 of the next section. Here we state the explicit functional equation for $D^r_{a,N}(s)$. Namely, by (3.7) we have, with $k = r+1/2$,

$$\begin{aligned} D^r_{a,N}(r+1-2s) &= N^{2s-r-1}\frac{\Gamma(s)\pi^{k-2s}}{\Gamma(k-s)} i^{-r} \sum_{0 \ne n \in \mathbf{Z}} \mathbf{e}(an/N)n^r|n|^{-2s} \\ &= N^{2s-r-1}\frac{\Gamma(s)\pi^{k-2s}}{\Gamma(k-s)} i^{-r} \sum_{b=1}^{N} \mathbf{e}(ab/N)D^r_{b,N}(2s-r). \end{aligned} \tag{3.15}$$

From (2.1b, f) we obtain $\Gamma(1/2-t) = \pi^{1/2}(-2)^t \prod_{\nu=1}^{t}(2\nu-1)^{-1}$ for $0 \le t \in \mathbf{Z}$. Therefore

$$\begin{aligned} &D^r_{a,N}(r+1-2m) \\ &= 2\cdot(2m-1-r)!N^{2m-r-1}(2\pi i)^{r-2m}\sum_{b=1}^{N}\mathbf{e}(ab/N)D^r_{b,N}(2m-r). \end{aligned} \tag{3.16}$$

Thus $D^r_{a,N}(r+1-2m)$ is a $\mathbf{Q}_{\mathrm{ab}}$-linear combination of $(2\pi i)^{r-2m}D^r_{b,N}(2m-r)$ for $0 \le b < N$. In Theorem 4.7 we will present a simpler way of expressing $D^r_{a,N}(r+1-2m)$.

CHAPTER II

CRITICAL VALUES OF DIRICHLET L-FUNCTIONS

4. The values of elementary Dirichlet series at integers

4.1. We first introduce the Euler numbers and the Euler polynomials not just in the classical sense, but in more generalized forms by putting

$$\frac{(1+c)e^z}{e^{2z}+c} = \sum_{n=0}^{\infty} \frac{E_{c,n}}{n!} z^n, \tag{4.1}$$

$$\frac{(1+c)e^{tz}}{e^z+c} = \sum_{n=0}^{\infty} \frac{E_{c,n}(t)}{n!} z^n. \tag{4.2}$$

Here $c = -\mathbf{e}(\alpha)$ with $\alpha \in \mathbf{R}, \notin \mathbf{Z}$. Thus $c \neq -1$. The standard **Euler numbers** and the **Euler polynomials** are $E_{1,n}$ and $E_{1,n}(t)$, that is, the symbols in the case $\alpha = 1/2$ and $c = 1$. Though they are usually written E_n and $E_n(t)$, we include 1 in the subscript for the purpose of distinguishing them from the Eisenstein series that we will introduce later. Thus $E_{c,n}$ and $E_{c,n}(t)$ may be called the **generalized Euler numbers** and the **generalized Euler polynomials.** To make our formulas short, put $b = -(1+c)^{-1}$. Then we have

$$E_{c,n} = 2^n E_{c,n}(1/2), \tag{4.3a}$$

$$E_{c,n}(t) = t^n + b\sum_{k=0}^{n-1}\binom{n}{k} E_{c,k}(t), \quad b = -(1+c)^{-1}, \tag{4.3b}$$

$$(d/dt)E_{c,n}(t) = nE_{c,n-1}(t) \qquad (n > 0), \tag{4.3c}$$

$$E_{c,n}(t+1) + cE_{c,n}(t) = (1+c)t^n, \tag{4.3d}$$

$$E_{1,2n+1} = 0, \tag{4.3e}$$

$$E_{c,n}(1-t) = (-1)^n E_{c^{-1},n}(t), \tag{4.3f}$$

$$E_{c^{-1},n}(0) = (-1)^{n+1} cE_{c,n}(0). \tag{4.3g}$$

All formulas except (4.3c) are valid for $n \geq 0$. Formula (4.3a) can be obtained by substituting $(1/2, 2z)$ for (t, z) in (4.2). Since $-b(e^z+c) = 1 - b\sum_{n=1}^{\infty} z^n/n!$, (4.2) can be written

$$\sum_{n=0}^{\infty} \frac{t^n z^n}{n!} = e^{tz} = \left\{1 - b\sum_{n=1}^{\infty} \frac{z^n}{n!}\right\} \sum_{n=0}^{\infty} \frac{E_{c,n}(t) z^n}{n!},$$

which produces (4.3b). Applying d/dt to (4.2), we obtain (4.3c). From (4.2) we obtain $\sum_{n=0}^{\infty}\{E_{c,n}(t+1)+cE_{c,n}(t)\}z^n/n! = (1+c)e^{tz}$, which gives (4.3d). We will prove (4.3e) and (4.3f) in the Remark after the following theorem. Combining (4.3d) with (4.3f), we obtain (4.3g). Clearly $E_{c,0}(t) = 1$. Using (4.3b), we can easily verify that $E_{c,n}(t)$ is a polynomial in t of degree n whose coefficients are polynomials in b with coefficients in $\mathbf{Z}$. For example,

$$\text{(4.3h)} \qquad E_{c,0}(t) = 1, \quad E_{c,1}(t) = t+b, \quad E_{c,2}(t) = t^2+2bt+2b^2+b,$$
$$E_{c,3}(t) = t^3+3bt^2+(6b^2+3b)t+6b^3+6b^2+b, \ \ b = -(1+c)^{-1}.$$

The significance of $E_{c,n}(t)$ is that it gives the value of the infinite series $F_{c,n}(t)$ (with $c = -\mathbf{e}(\alpha)$ as above) defined by

$$\text{(4.4)} \qquad F_{c,n}(t) = \sum_{h\in\mathbf{Z}}(h+\alpha)^{-n-1}\mathbf{e}((h+\alpha)t) \qquad (0 \le n \in \mathbf{Z},\ t \in \mathbf{R}).$$

The infinite sum on the right-hand side depends only on α (mod $\mathbf{Z}$), and so the notation $F_{c,n}$ is meaningful. Also, the sum is clearly convergent for $n > 0$. If $n = 0$, we understand that $\sum_{h\in\mathbf{Z}}$ means $\lim_{m\to\infty}\sum_{|h|\le m}$, which is indeed meaningful, as will be shown below. We have

$$\text{(4.4a)} \qquad F_{c,n}(t+m) = \mathbf{e}(m\alpha)F_{c,n}(t) \quad \text{if} \quad m \in \mathbf{Z},$$

$$\text{(4.4b)} \qquad F_{c,n}(1-t) = (-1)^n cF_{c^{-1},n}(t).$$

Formula (4.4a) is obvious. Replacing h by $-h$ in (4.4), we obtain (4.4b). Taking $\alpha = q/N$ with a positive integer N and an integer q such that $q \notin N\mathbf{Z}$, we easily see that

$$\text{(4.4c)} \qquad F_{c,k-1}(Nt) = N^k \sum_{m\in q+N\mathbf{Z}} m^{-k}\mathbf{e}(mt) \qquad (c = -\mathbf{e}(q/N),\ 0 < k \in \mathbf{Z}).$$

Theorem 4.2. *For $0 \le n \in \mathbf{Z}$ and $0 < t < 1$ we have*

$$\text{(4.5)} \qquad E_{c,n}(t) = (1+c^{-1})n!(2\pi i)^{-n-1}F_{c,n}(t).$$

This is true even for $0 \le t \le 1$ if $n > 0$. Moreover,

$$\text{(4.5a)} \qquad (1+c^{-1})(2\pi i)^{-1}F_{c,0}(0) = (1-c^{-1})/2.$$

Remark. These formulas combined with (4.4a) determine $F_{n,c}(t)$ for every $t \in \mathbf{R}$. Notice that $E_{c,0}(0) = 1$, which is different from (4.5a). Putting $t = \alpha = 1/2$ in (4.4) and (4.5), we obtain (4.3e). Formula (4.3f) follows from (4.4b) and (4.5).

PROOF. We obtain (4.5a) directly from (2.18). We prove (4.5) first in the case $n > 0$ and $0 \le t \le 1$ by taking the contour integral

$$(*) \qquad \int_S f(z)dz, \qquad f(z) = \frac{e^{tz}}{z^{n+1}(e^z+c)},$$

where S is a square, with center at the origin, whose vertices are $A \pm iA$ and $-A \pm iA$, $A = 2N\pi$ with $0 < N \in \mathbf{Z}$. The poles of f in $\mathbf{C}$ are 0 and $2\pi i(h+\alpha)$ with $h \in \mathbf{Z}$. The residue of f at $2\pi i(h+\alpha)$ is easily seen to be $-c^{-1}\mathbf{e}(t(h+\alpha))(2\pi i(h+\alpha))^{-n-1}$. From (4.2) we see that the residue of f at 0 is $(1+c)^{-1}E_{c,n}(t)/n!$. Thus by the theorem of residues, the integral of (4.5) equals $2\pi i$ times

$$(1+c)^{-1}E_{c,n}(t)/n! - c^{-1}(2\pi i)^{-n-1}\sum_{h}(h+\alpha)^{-n-1}\mathbf{e}((h+\alpha)t),$$

where h runs over the integers such that $|h+\alpha| < N$. To make an estimate of the integral on the sides of S, put $z = x + iy$. If $x = A$, then $|e^{tz}| = e^{tA}$ and $|e^z + c| \geq e^A - 1$, and so $|e^{tz}/(e^z+c)| \leq e^A/(e^A-1)$. If $x = -A$, then $|e^{tz}| = e^{-tA} \leq 1$ and $|e^z + c| \geq 1 - e^{-A}$, and so $|e^{tz}/(e^z+c)| \leq e^A/(e^A-1)$. Next suppose $y = \pm iA$. If $e^x > 2$, then $|e^{tz}| = e^{tx} \leq e^x$ and $|e^z + c| \geq e^x - 1$, and so $|e^{tz}/(e^z+c)| \leq e^x/(e^x-1) = 1/(1-e^{-x}) \leq 2$. If $e^x \leq 2$, then $|e^{tz}| = e^{tx} \leq 2$ and $|e^z + c| = |e^x - \mathbf{e}(\alpha)| \geq B$ with a positive constant B depending only on α, and so $|e^{tz}/(e^z+c)| \leq 2/B$. Therefore, because of the factor z^{-n-1}, we see that the integral of $(*)$ tends to 0 as $N \to \infty$. This proves (4.5) for $n > 0$ with $0 \leq t \leq 1$. To prove the case $n = 0$, we need the following lemma.

Lemma 4.3. (i) *Let $\{a_n\}_{n=1}^{\infty}$ be an increasing sequence of positive numbers such that $\lim_{n\to\infty} a_n = \infty$ and let T be a compact subset of $\mathbf{R}$ such that $T \cap \mathbf{Z} = \emptyset$. Then the series $\sum_{n=1}^{\infty} a_n^{-s}\mathbf{e}((\alpha+n)t)$ and $\sum_{n=1}^{\infty} a_n^{-s}\mathbf{e}((\alpha-n)t)$ are uniformly convergent for a fixed $\alpha \in \mathbf{C}$, $t \in T$, and $s > \sigma$ with any positive constant σ.*

(ii) *Let χ be a $\mathbf{C}$-valued function on $\mathbf{Z}/N\mathbf{Z}$ such that $\sum_{a=1}^{N}\chi(a) = 0$, and let $\{a_n\}_{n=1}^{\infty}$ be as in* (i). *Then $\sum_{n=1}^{\infty} a_n^{-s}\chi(n)$ is uniformly convergent for $s > \sigma$ with any positive constant σ.*

PROOF. To prove (i), put $M = \mathrm{Max}_{t\in T} 2/|1-\mathbf{e}(t)|$, $\omega = \mathbf{e}(t)$, and $\gamma_n = \sum_{h=1}^{n}\omega^h$. Then $|\gamma_n| = |(\omega - \omega^{n+1})/(1-\omega)| \leq M$ if $t \in T$, and so for $1 < m \leq n$ we have

$$\begin{aligned}\sum_{h=m}^{n} a_h^{-s}\mathbf{e}(ht) &= \sum_{h=m}^{n}(\gamma_h - \gamma_{h-1})a_h^{-s} \\ &= \sum_{h=m}^{n-1}\gamma_h\{a_h^{-s} - a_{h+1}^{-s}\} + \gamma_n a_n^{-s} - \gamma_{m-1}a_m^{-s}.\end{aligned}$$

Thus $\left|M^{-1}\sum_{h=m}^{n} a_h^{-s}\mathbf{e}(ht)\right| \leq a_m^{-s} + a_n^{-s} + \sum_{h=m}^{n-1}(a_h^{-s} - a_{h+1}^{-s}) \leq 2a_m^{-s}$, which proves the desired uniform convergence of $\sum_{n=1}^{\infty} a_n^{-s}\mathbf{e}((\alpha+n)t)$, as $|\mathbf{e}(\alpha t)|$ is bounded for $t \in T$. The case with $\alpha - n$ in place of $\alpha + n$ can be handled in the same way. To prove (ii), we replace γ_n in the above proof by $\gamma_n' = \sum_{h=1}^{n}\chi(h)$. We easily see that the $|\gamma_n'|$ are bounded for all n, and we obtain the desired fact by the same technique.

To complete the proof of Theorem 4.2, we consider the series

$$\sum_{n=1}^{\infty}(n+\alpha)^{-2}\mathbf{e}((n+\alpha)t)+\alpha^{-2}\mathbf{e}(\alpha t)+\sum_{n=1}^{\infty}(n-\alpha)^{-2}\mathbf{e}((\alpha-n)t),$$

$$\sum_{n=1}^{\infty}(n+\alpha)^{-1}\mathbf{e}((n+\alpha)t)+\alpha^{-1}\mathbf{e}(\alpha t)-\sum_{n=1}^{\infty}(n-\alpha)^{-1}\mathbf{e}((\alpha-n)t).$$

Termwise application of d/dt to the first series produces $2\pi i$ times the second series. Since $(d/dt)E_{c,1}(t) = E_{c,0}(t)$, we thus obtain (4.5) for $E_{c,0}(t)$, provided such termwise differentiation is valid, which is indeed the case, as Lemma 4.3 shows that these series are uniformly convergent for $b \le t \le b'$ with constants b and b' such that $0 < b < b' < 1$. As for s, we simply take it to be 1 or 2. (Actually, the standard theorem on termwise differentiation requires the uniform convergence only for the latter series.)

4.4. Taking $t = 0$ in (4.5) with $n > 0$ and comparing the result with (2.16), we find that

$$P_{n+1}(x) = (x-1)^n E_{-x,n}(0) \qquad (n > 0), \tag{4.6}$$

where x is an indeterminate. We can make it valid even when $n = 0$ by putting $P_1(x) = 1$. This combined with (4.3g) resp. (4.3b) proves (2.19) resp. (2.20).

4.5. The right-hand side of (4.4) has $\alpha^{-n-1}\mathbf{e}(\alpha t)$ as the term for $h = 0$, which makes the sum meaningless if $\alpha = 0$. However, removing that term, for $\alpha = 0$ we have $\sum_{0\neq h\in\mathbf{Z}} h^{-n-1}\mathbf{e}(ht)$. We will now show that this sum can be handled by introducing the **Bernoulli numbers** B_n and **Bernoulli polynomials** $B_n(t)$ for $0 \le n \in \mathbf{Z}$ as follows:

$$\frac{z}{e^z-1} = \sum_{n=0}^{\infty}\frac{B_n}{n!}z^n, \tag{4.7a}$$

$$\frac{ze^{tz}}{e^z-1} = \sum_{n=0}^{\infty}\frac{B_n(t)}{n!}z^n. \tag{4.7b}$$

We have, for every $n \ge 0$ unless stated otherwise,

$$B_n = B_n(0), \tag{4.8a}$$

$$\sum_{k=0}^{n-1}\binom{n}{k}B_k(t) = nt^{n-1} \qquad (n > 0), \tag{4.8b}$$

$$(d/dt)B_n(t) = nB_{n-1}(t) \qquad (n > 0), \tag{4.8c}$$

$$B_n(t+1) - B_n(t) = nt^{n-1}, \tag{4.8d}$$

$$B_{2m+1} = 0 \qquad (0 < m \in \mathbf{Z}), \tag{4.8e}$$

$$B_n(1-t) = (-1)^n B_n(t). \tag{4.8f}$$

These are well known, and can be proved in the same way as for (4.3a–f). Also, $B_n(t)$ is a polynomival in t of degree n with rational coefficients. For example,

(4.8f) $B_0(t) = 1,\ B_1(t) = t - \frac{1}{2},\ B_2(t) = t^2 - t + \frac{1}{6},\ B_3(t) = t^3 - \frac{3}{2}t^2 + \frac{1}{2}t.$

Theorem 4.6. *For $1 \le n \in \mathbf{Z}$ and $0 < t < 1$ we have*

$$B_n(t) = -n!(2\pi i)^{-n} \sum_{0 \ne h \in \mathbf{Z}} h^{-n}\mathbf{e}(ht), \tag{4.9}$$

where $\sum_{0\ne h\in\mathbf{Z}}$ means $\lim_{m\to\infty}\sum_{0<|h|\le m}$ when $n = 1$. This formula is valid even for $0 \le t \le 1$ if $n > 1$.

PROOF. This can be proved in the same manner as for Theorem 4.2. We consider

$$\int_S f(z)dz, \qquad f(z) = \frac{e^{tz}}{z^n(e^z - 1)},$$

where S is a square, with center at the origin, whose vertices are $A \pm iA$ and $-A \pm iA$, $A = (2N+1)\pi$ with $0 < N \in \mathbf{Z}$. The poles of f in $\mathbf{C}$ are $2\pi i h$ with $h \in \mathbf{Z}$. If $h \ne 0$, the residue of f at $2\pi i h$ is easily seen to be $(2\pi i h)^{-n}\mathbf{e}(th)$. From (4.7b) we see that the residue of f at 0 is $B_n(t)/n!$. Using the theorem of residues and making N large, we obtain (4.9) for $n > 1$ and $0 \le t \le 1$. Since $B_2'(t) = 2B_1(t)$, we obtain (4.9) for $B_1(t)$ by applying d/dt to the formula for $B_2(t)$, as we can justify termwise differentiation when $0 < t < 1$, by virtue of Lemma 4.3.

Now the main theorem on the values of $D^r_{a,N}(s)$ at integers can be stated as follows:

Theorem 4.7. *Given $0 < N \in \mathbf{Z}$, $a \in \mathbf{Z}$, and $r = 0$ or 1, define $D^r_{a,N}(s)$ by (3.11)* and put $\xi = \mathbf{e}(a/N)$. Then for $0 < m \in \mathbf{Z}$ the following assertions hold:

(i) *The quantities $(2\pi i)^{-2m}D^0_{a,N}(2m)$ and $(2\pi i)^{1-2m}D^1_{a,N}(2m-1)$ for $0 < a < N$ are numbers of $\mathbf{Q}(\xi)$ given as follows:*

$$(2m-1)!N^{2m}(2\pi i)^{-2m}D^0_{a,N}(2m) = \frac{\xi}{\xi - 1}E_{-\xi,2m-1}(0) = \frac{\xi P_{2m}(\xi)}{(\xi-1)^{2m}}, \tag{4.10a}$$

$$(2m-2)!N^{2m-1}(2\pi i)^{1-2m}D^1_{a,N}(2m-1) \tag{4.10b}$$
$$= \begin{cases} \dfrac{\xi}{\xi-1}E_{-\xi,2m-2}(0) = \dfrac{\xi P_{2m-1}(\xi)}{(\xi-1)^{2m-1}} & (m > 1), \\[2ex] \dfrac{\xi+1}{2(\xi-1)} & (m = 1). \end{cases}$$

(ii) *If $\sigma \in \mathrm{Gal}(\mathbf{Q}(\zeta)/\mathbf{Q})$ and $\mathbf{e}(1/N)^\sigma = \mathbf{e}(q/N)$ with $q \in \mathbf{Z}$, then*

$$\{(2\pi i)^{r-2m}D^r_{a,N}(2m-r)\}^\sigma = (2\pi i)^{r-2m}D^r_{qa,N}(2m-r). \tag{4.11}$$

(iii) *In the case $a = 0$ we have*

$$-(2m)!(2\pi i)^{-2m}N^{2m}D^0_{0,N}(2m) = B_{2m}. \tag{4.12}$$

(iv) *The quantity* $D^r_{a,N}(r+1-2m)$ *for* $0<a<N$ *is a rational number given by*

$$D^r_{a,N}(r+1-2m) = \frac{2}{r-2m} N^{2m-r-1} B_{2m-r}(a/N). \tag{4.13}$$

This is true even for $a=0$ *if* $2m-r>1$.

Remark. From (4.12) we see that $(-1)^{m+1}B_{2m} > 0$.

PROOF. The first two formulas are already given in (3.9), as $D^r_{a,N}(2m-r) = M^{2m-r}_{a,N}(0)$ and $P_{n+1}(x) = (x-1)^n E_{-x,n}(0)$; see (3.12) and (4.6). We can also prove them by taking $\alpha = a/N$ and $t=0$ in (4.5), excluding the case involving $D^1_{a,N}(1)$. Formula (4.12) follows from (4.9) with $t=0$. Assertion (ii) can easily be seen from (4.10a, b). Putting $t=a/N$ in (4.9) and comparing the result with (3.16), we obtain (4.13).

4.8. For $1<N\in\mathbf{Z}$ let χ be a primitive or an imprimitive Dirichlet character modulo N. This means that χ is a homomorphism $(\mathbf{Z}/N\mathbf{Z})^\times \to \mathbf{C}^\times$ which can be trivial. We put $\chi(a)=0$ for a not prime to N. We define, as usual, the Dirichlet L-function $L(s,\chi)$ by

$$L(s,\chi) = \sum_{n=1}^{\infty} \chi(n)n^{-s}. \tag{4.14}$$

The right-hand side is absolutely convergent for $\mathrm{Re}(s)>1$, and convergent even for $\mathrm{Re}(s)>0$ provided χ is nontrivial, by virtue of Lemma 4.3 (ii). Suppose $\chi(-1)=(-1)^r$ with $r=0$ or 1; then

$$2L(s,\chi) = \sum_{a=1}^{N} \chi(a) D^r_{a,N}(s) = \sum_{a=1}^{N} \chi(a) M^{-r}_{a,N}\big((s+r)/2\big), \tag{4.15}$$

and so $\Gamma\big((s+r)/2\big)L(s,\chi)$ can be continued as a meromorphic function to the whole s-plane with possible poles at $s=0$ and $s=1$, which occur only when χ is trivial.

Theorem 4.9. *Let* χ *and* N *be as in §4.8; let* k *be a positive integer such that* $\chi(-1)=(-1)^k$, $m=[(N-1)/2]$, *and* $\zeta=\mathbf{e}(1/N)$. *Then*

$$N^k(k-1)!(2\pi i)^{-k}L(k,\chi) = \sum_{a=1}^{m} \chi(a)\frac{\zeta^a E_{-\zeta^a,k-1}(0)}{\zeta^a-1} \tag{4.16a}$$

$$= \sum_{a=1}^{m} \chi(a)\frac{\zeta^a P_k(\zeta^a)}{(\zeta^a-1)^k} \qquad (1<k\in\mathbf{Z}),$$

$$L(1,\chi) = \frac{\pi i}{N}\sum_{a=1}^{m}\chi(a)\frac{\zeta^a+1}{\zeta^a-1} \quad \text{if} \quad \chi(-1)=-1. \tag{4.16b}$$

Moreover, let μ_d denote the primitive character modulo d, where d is 3 or 4. Then we have, with $\zeta=\mathbf{e}(1/d)$,

$$\text{(4.17a)}\quad d^k(k-1)!(2\pi i)^{-k}L(k,\,\mu_d) = \frac{\zeta P_k(\zeta)}{(\zeta-1)^k} = \frac{\zeta E_{-\zeta,k-1}(0)}{\zeta-1} \quad (0<k-1\in 2\mathbf{Z}),$$

$$\text{(4.17b)}\quad dL(1,\,\mu_d) = \pi i(\zeta+1)/(\zeta-1).$$

PROOF. Since $2L(k+2s,\,\chi) = \sum_{a=1}^{N-1}\chi(a)M^k_{a,N}(s)$ if $\chi(-1)=(-1)^k$, from (3.9) and (4.6) we obtain, for $k>1$,

$$2N^k(k-1)!(2\pi i)^{-k}L(k,\,\chi) = \sum_{a=1}^{N-1}\chi(a)\frac{\zeta^a E_{-\zeta^a,k-1}(0)}{\zeta^a-1}.$$

Now $N=2m+1$ or $2m+2$, and so the last sum can be written

$$\sum_{a=1}^{m}\left\{\chi(a)\frac{E_{-\zeta^a,k-1}(0)}{1-\zeta^{-a}} + \chi(N-a)\frac{E_{-\zeta^{-a},k-1}(0)}{1-\zeta^a}\right\},$$

as $\chi(m+1)=0$ if N is even. Using (4.3g), we obtain the part of (4.16a) involving $E_{*,*}$, which combined with (4.6) gives the remaining part. We can prove (4.16b) in the same manner by means of (3.9) with $\nu=1$. Formulas (4.17a, b) are mere special cases of (4.16a, b).

The formulas of the above theorem are different from the classical formulas for $L(k,\,\chi)$, which we will present in Theorem 4.12 below.

4.10. Take $N = 2, 3, 4$, or 6 and $a = 0$ or 1. Then we easily see that $D^0_{0,N}(s) = 2N^{-s}\zeta(s)$ and

$$\text{(4.18a)}\quad D^0_{1,2}(s) = 2(1-2^{-s})\zeta(s),$$

$$\text{(4.18b)}\quad D^0_{1,3}(s) = (1-3^{-s})\zeta(s),$$

$$\text{(4.18c)}\quad D^0_{1,4}(s) = (1-2^{-s})\zeta(s),$$

$$\text{(4.18d)}\quad D^0_{1,6}(s) = (1-2^{-s})(1-3^{-s})\zeta(s).$$

Thus $\zeta(2m)$ and $\zeta(1-2m)$ can be obtained from (4.10a) and (4.13). For instance, for $0<k\in 2\mathbf{Z}$ we have

$$\text{(4.19a)}\quad (k/2)\zeta(1-k) = \frac{2^{k-1}}{2^k-2}B_k(1/2) = \frac{3^{k-1}}{3^{k-1}-1}B_k(1/3)$$
$$= \frac{4^{k-1}}{2^{k-1}-1}B_k(1/4) = \frac{-6^{k-1}}{(2^{k-1}-1)(3^{k-1}-1)}B_k(1/6),$$

$$\text{(4.19b)}\quad \frac{(k-1)!\zeta(k)}{(2\pi i)^k} = \frac{-P_k(-1)}{2^{k+1}(2^k-1)} = \frac{\omega^2P_k(\omega^2)}{(3^k-1)(\omega^2-1)^k}$$
$$= \frac{iP_k(i)}{(4^k-2^k)(i-1)^k} = \frac{\omega P_k(\omega)}{(2^k-1)(3^k-1)(\omega-1)^k},$$

where $\omega = \mathbf{e}(1/6)$. Comparing (4.19b) with (4.12), we obtain the part of the following formula concerning even k.

$$\text{(4.19c)}\quad \frac{2P_k(i)}{(i-1)^k} = \begin{cases} k^{-1}i(4^k-2^k)B_k & (0<k\in 2\mathbf{Z}),\\ -E_{1,\,k-1} & (0<k-1\in 2\mathbf{Z}).\end{cases}$$

The part for odd k follows from (4.17a) and (4.30b) below.

4.11. Let us now treat the values of $L(k, \chi)$ first in the traditional way and then in several novel ways. We take a primitive Dirichlet character χ modulo d, and define the **Gauss sum** $G(\chi)$ by

$$G(\chi) = \sum_{a=1}^{d} \chi(a)\mathbf{e}(a/d). \tag{4.20}$$

The following properties of $G(\chi)$ are well known:

$$\sum_{a=1}^{d} \chi(a)\mathbf{e}(ab/d) = \overline{\chi}(b)G(\chi) \quad \text{for every } b \in \mathbf{Z}, \tag{4.21}$$

$$G(\chi)G(\overline{\chi}) = \chi(-1)d, \quad \overline{G(\chi)} = \chi(-1)G(\overline{\chi}), \quad |G(\chi)|^2 = d. \tag{4.22}$$

See [S71, Lemma 3.63], for example. Here and throughout the rest of this section d is a positive integer

From (3.7) or (3.15) we can easily derive the functional equation for $L(s, \chi)$. Namely, Let χ be a primitive character modulo d such that $\chi(-1) = (-1)^r$ with $r = 0$ or 1. Put

$$R(s, \chi) = (d/\pi)^{s/2}\Gamma\big((s+r)/2\big)L(s, \chi). \tag{4.23}$$

Then

$$R(s, \chi) = W(\chi)R(1-s, \overline{\chi}) \quad \text{with} \quad W(\chi) = i^{-r}d^{-1/2}G(\chi). \tag{4.24}$$

Indeed, from (3.7) and (4.21) we obtain, with $k = r + 1/2$,

$$\pi^{s-k}\Gamma(k-s)\sum_{a=1}^{d} \overline{\chi}(a)M_{a,d}^{-r}(k-s) = d^{2s-r-1}\pi^{-s}\Gamma(s)i^{-r}G(\overline{\chi})\sum_{b=1}^{d} \chi(b)M_{\overline{b},d}^{-r}(s).$$

Substituting $(s+r)/2$ for s and employing (4.22), we obtain (4.24).

Theorem 4.12. (i) *Let $\zeta(s)$ denote the Riemann zeta function. Then for $0 < n \in 2\mathbf{Z}$ we have*

$$2 \cdot n!(2\pi i)^{-n}\zeta(n) = n\zeta(1-n) = -B_n. \tag{4.25}$$

(ii) *Let χ be a nontrivial primitive Dirichlet character modulo d, and let k be a positive integer such that $\chi(-1) = (-1)^k$. Then*

$$2 \cdot k!(2\pi i)^{-k}G(\overline{\chi})L(k, \chi) = kd^{1-k}L(1-k, \overline{\chi}) = -\sum_{a=1}^{d-1} \overline{\chi}(a)B_k(a/d). \tag{4.26}$$

In particular, if $k = 1$ and $\chi(-1) = -1$, then

$$(\pi i)^{-1}G(\chi)L(1, \overline{\chi}) = L(0, \chi) = -\sum_{a=1}^{d-1} \chi(a)a/d. \tag{4.27}$$

PROOF. Formula (4.25) can be obtained from (4.8a), (4.12), and (4.13) by taking $N = 1$ and $r = a = 0$, as $D_{0,1}^{0}(s) = 2\zeta(s)$. As for (4.26), by (4.21) and (4.9) we have

$$2G(\overline{\chi})L(k,\chi) = \sum_{0\neq h\in\mathbf{Z}} h^{-k}G(\overline{\chi})\chi(h)$$

$$= \sum_{0\neq h\in\mathbf{Z}}\sum_{a=1}^{d} h^{-k}\overline{\chi}(a)\mathbf{e}(ha/d) = -\frac{(2\pi i)^k}{k!}\sum_{a=1}^{d}\overline{\chi}(a)B_k(a/d).$$

If $k=1$, we need to show that $2L(1,\chi) = \lim_{m\to\infty}\sum_{0\neq|h|\leq m}\chi(h)h^{-1}$, which follows from the uniform convergence of $\sum_{h=1}^{\infty}\chi(h)h^{-s}$ for $s>1/2$, as guaranteed by Lemma 4.3 (ii). The value $L(1-k,\overline{\chi})$ can be obtained by considering $\sum_{a=1}^{d}\overline{\chi}(a)D_{a,d}^{r}(r+1-2m)$ and employing (4.13) and (4.15) with $N=d$ and $k=2m-r$. It also follows from (4.24). Formula (4.27) is merely a special case of (4.26), as $B_1(t)=t-1/2$ and $\sum_{a=1}^{d-1}\overline{\chi}(a)=0$.

Let us insert here a few comments. The first one is historical. Formula (4.25) is due to Euler. Dirichlet proved (4.27) and even found a formula for $L(1,\chi)$ when $\chi(-1)=1$. As for (4.26), Hecke stated and proved it in [He40], and he is perhaps the first person who did so, though the fact was possibly known to other number-theorists, as (4.9) had been known since much earlier periods.

In any case, (4.26) is the well known standard formula for $L(k,\chi)$, but there is no reason for accepting it as the best or most important result. Indeed, we have a clear-cut formula (4.13), which we can view as more basic than (4.26). If the values of Dirichlet series are our main interest, the series $D_{a,N}^{r}(s)$ and the infinite sum of (4.4c) are most natural objects of study. In the same spirit we will present various new formulas for $L(k,\chi)$ which are different from (4.26) and are derived by the idea expressed by (4.4c).

We first prove

Lemma 4.13. (i) *Let χ be a nontrivial primitive Dirichlet character modulo d, and let k be a positive integer such that $\chi(-1)=(-1)^k$. Further let $q=[(d-1)/2]$. Then*

$$\sum_{a=1}^{d-1}\chi(a)B_n(a/d) = \begin{cases} 2\sum_{a=1}^{q}\chi(a)B_n(a/d) & \text{if } n=k,\\ 0 & \text{if } n=k-1.\end{cases} \tag{4.28}$$

(ii) *In the setting of (i), suppose d is odd. Then*

$$\sum_{a=1}^{d-1}(-1)^a\chi(a)E_{1,n}(a/d) = \begin{cases} 2\sum_{a=1}^{q}(-1)^a\chi(a)E_{1,n}(a/d) & \text{if } n=k-1,\\ 0 & \text{if } n=k.\end{cases} \tag{4.29}$$

(iii) *In the setting of (i), suppose $d=4d_0$ with $1<d_0\in\mathbf{Z}$. Then*

$$\sum_{a=1}^{q}\chi(a)E_{1,n}(2a/d) = \begin{cases} 2\sum_{a=1}^{d_0-1}\chi(a)E_{1,n}(2a/d) & \text{if } n=k-1,\\ 0 & \text{if } n=k.\end{cases} \tag{4.30}$$

(iv) *For* χ, d, *and* q *as in (i), suppose* $\chi(-1) = 1$. *Then*

$$\sum_{a=1}^{q} \chi(a) = 0, \tag{4.31a}$$

$$\sum_{a=1}^{d-1} \chi(a)a = 0, \tag{4.31b}$$

$$\sum_{a=1}^{d-1} (-1)^a \chi(a) = 0 \quad \text{if} \quad d \notin 2\mathbf{Z}, \tag{4.31c}$$

$$\sum_{a=1}^{d-1} \chi(a)a^3 = (3d/2) \sum_{a=1}^{d-1} \chi(a)a^2. \tag{4.31d}$$

PROOF. We have $d = 2q + 1$ or $d = 2q + 2$ according as d is odd or even. If d is even, then $4|d$, and so $\chi(q + 1) = 0$. By (4.8f) we have $B_n\big((d - a)/d\big) = (-1)^n B_n(a/d)$, and so

$$\begin{aligned}\sum_{a=1}^{d-1} \chi(a)B_n(a/d) &= \sum_{a=1}^{q} \chi(a)B_n(a/d) + \sum_{a=1}^{q} \chi(d - a)B_n\big((d - a)/d\big)\\ &= \sum_{a=1}^{q} \chi(a)\{B_n(a/d) + (-1)^{k+n}B_n(a/d)\}.\end{aligned}$$

This proves (4.28). We can similarly prove (4.29); the only difference is that we have $(-1)^{d+k+n}$ instead of $(-1)^{k+n}$. If $d = 4d_0 > 4$ as in (iii), we have $q = 2d_0 - 1$, and so

$$\sum_{a=1}^{q} \chi(a)E_{1,n}(2a/d) = \sum_{a=1}^{d_0-1} \Big\{\chi(a)E_{1,n}(2a/d) + \chi(2d_0 - a)E_{1,n}\big(2(2d_0 - a)/d\big)\Big\},$$

as $\chi(d_0) = 0$. Since $2d_0a - 2d_0 \in d\mathbf{Z}$ for odd a, we have $\chi(2d_0 - a) = \chi(2d_0a - a) = \chi(-a)\chi(1 - 2d_0)$. Thus by (4.3f) the last sum equals

$$\{1 + (-1)^{n+k}\chi(1 - 2d_0)\} \sum_{a=1}^{d_0-1} \chi(a)E_{1,n}(2a/d).$$

By Lemma 1.12 we have $\chi(1 - 2d_0) = -1$. Thus we obtain (4.30). Finally suppose $\chi(-1) = 1$. Then $\chi(d - a) = \chi(a)$, and so

$$0 = \sum_{a=1}^{d-1} \chi(a) = \sum_{a=1}^{q} \chi(a) + \sum_{a=1}^{q} \chi(d - a) = 2\sum_{a=1}^{q} \chi(a),$$

which proves (4.31a). (Notice that $\chi(q + 1) = 0$ if d is even.) Therefore

$$\sum_{a=1}^{d-1} \chi(a)a = \sum_{a=1}^{q} \chi(a)a + \sum_{a=1}^{q} \chi(d - a)(d - a) = d\sum_{a=1}^{q} \chi(a) = 0,$$

which is (4.31b). Similarly, employing (4.31a), we find that

$$\sum_{a=1}^{d-1}\chi(a)a^2=\sum_{a=1}^{q}\chi(a)\{a^2+(d-a)^2\}=2\sum_{a=1}^{q}\chi(a)(a^2-da),$$

$$\sum_{a=1}^{d-1}\chi(a)a^3=\sum_{a=1}^{q}\chi(a)\{a^3+(d-a)^3\}=3d\sum_{a=1}^{q}\chi(a)(a^2-da).$$

Formula (4.31d) follows from these two equalities. If d is odd,

$$\sum_{a=1}^{d-1}(-1)^a\chi(a)=\sum_{b=1}^{d-1}(-1)^{d-b}\chi(d-b)=-\sum_{b=1}^{d-1}(-1)^b\chi(b),$$

from which we obtain (4.31c).

We now state the principal results of this section, among which (4.32), (4.34), and (4.35) are most essential.

Theorem 4.14. (i) *Let χ be a nontrivial primitive Dirichlet character modulo d, and let k be a positive integer such that $\chi(-1)=(-1)^k$. Further let $q=[(d-1)/2]$. Then*

$$(4.32)\qquad (k-1)!(2\pi i)^{-k}G(\overline{\chi})L(k,\chi)=\frac{1}{2(2^k-\chi(2))}\sum_{a=1}^{q}\overline{\chi}(a)E_{1,k-1}(2a/d),$$

$$(4.33)\qquad k!(2\pi i)^{-k}G(\overline{\chi})L(k,\chi)=-\sum_{a=1}^{q}\overline{\chi}(a)B_k(a/d).$$

(ii) *Suppose in particular $d=4d_0$ with $1<d_0\in\mathbf{Z}$. Then*

$$(4.34)\qquad (k-1)!(\pi i)^{-k}G(\overline{\chi})L(k,\chi)=\sum_{a=1}^{d_0-1}\overline{\chi}(a)E_{1,k-1}(2a/d).$$

(iii) *If d is odd, then*

$$(4.35)\qquad (k-1)!(2\pi i)^{-k}G(\overline{\chi})L(k,\chi)=\frac{\chi(2)}{2(2^k-\chi(2))}\sum_{b=1}^{q}(-1)^b\overline{\chi}(b)E_{1,k-1}(b/d).$$

In particular, if $k=1$, $m=[q/2]$, and $n=[(q-1)/2]$, then we have

$$(4.36)\qquad (2\pi i)^{-1}G(\overline{\chi})L(1,\chi)$$
$$=\begin{cases}\dfrac{1}{\{1+\chi(2)\}\{2-\chi(2)\}}\displaystyle\sum_{a=1}^{m}\overline{\chi}(a) & \text{if } \chi(2)\neq-1,\\[2ex] \dfrac{1}{\{1-\overline{\chi}(2)\}\{2-\chi(2)\}}\displaystyle\sum_{c=0}^{n}\overline{\chi}(2c+1) & \text{if } \chi(2)\neq 1.\end{cases}$$

(iv) *If μ_d denotes the primitive character modulo d with $d=3$ or 4, then for $k=2m+1$ with $0\le m\in\mathbf{Z}$ we have*

$$(4.37)\qquad k!(-1)^m(2\pi)^{-k}\sqrt{3}\,L(k,\mu_3)=-B_k\big(\tfrac{1}{3}\big)=\frac{k}{2(2^k+1)}E_{1,2m}\big(\tfrac{2}{3}\big),$$

(4.38) $$k!(-4)^{-m}\pi^{-k}L(k,\,\mu_4) = -B_k(\tfrac{1}{4}) = 2^{-2k}kE_{1,\,2m}.$$

PROOF. In the proof of Lemma 4.13 we noted that $\chi(q+1)=0$ if $d \in 2\mathbf{Z}$. Therefore by (4.21) we have, for every $m \in \mathbf{Z}$,

$$G(\overline{\chi})\chi(m) = \sum_{a=1}^{q}\overline{\chi}(a)\mathbf{e}(am/d) + \sum_{a=1}^{q}\overline{\chi}(d-a)\mathbf{e}((d-a)m/d)$$
$$= \sum_{a=1}^{q}\overline{\chi}(a)\{\mathbf{e}(am/d) + (-1)^k\mathbf{e}(-am/d)\}.$$

Take $n=k-1$, $\alpha = 1/2$, and $t{=}2a/d$ with $1 \le a \le q$ in (4.4) and (4.5). Then

(*) $$E_{1,k-1}(2a/d) = 2(k-1)!(2\pi i)^{-k}2^k\sum_{m\ \mathrm{odd}} m^{-k}\mathbf{e}(ma/d).$$

On the other hand,

$$2G(\overline{\chi})(1-\chi(2)2^{-k})L(k,\,\chi) = G(\overline{\chi})\sum_{m\ \mathrm{odd}}\chi(m)m^{-k}$$
$$= \sum_{a=1}^{q}\overline{\chi}(a)\Big\{\sum_{m\ \mathrm{odd}} m^{-k}\mathbf{e}(am/d) + \sum_{m\ \mathrm{odd}}(-m)^{-k}\mathbf{e}(-am/d)\Big\}$$
$$= 2\sum_{a=1}^{q}\overline{\chi}(a)\sum_{m\ \mathrm{odd}} m^{-k}\mathbf{e}(am/d).$$

Combining this with (*), we obtain (4.32). We have to be careful about the case $k=1$, but Lemma 4.3 settles the technical point as explained in the proof of Theorem 4.12. Formula (4.33) follows immediately from (4.26) combined with (4.28).

If $d = 4d_0 > 4$, then $q = 2d_0 - 1$, and we have

$$\sum_{a=1}^{q}\overline{\chi}(a)E_{1,k-1}(2a/d) = \{1-\overline{\chi}(1-2d_0)\}\sum_{a=1}^{d_0-1}\overline{\chi}(a)E_{1,k-1}(2a/d),$$

as shown in the proof of Lemma 4.13. (Take $n=k-1$.) Since $(1-2d_0)^2-1 \in d\mathbf{Z}$, we have $\chi(1-2d_0) = \pm 1$. If $\chi(1-2d_0)=1$, then (4.32) shows that $L(k,\,\chi)=0$, a contradiction. Thus $\chi(1-2d_0) = -1$, and we obtain (4.34) from (4.32), as $\chi(2)=0$. (We already stated the equality $\chi(1-2d_0)=-1$ in Lemma 1.12 and gave an elementary proof.)

As for (iii), putting $m = [q/2]$ and $n = [(q-1)/2]$, we have clearly

$$\overline{\chi}(2)\sum_{a=1}^{q}\overline{\chi}(a)E_{1,k-1}(2a/d) = \sum_{a=1}^{m}\overline{\chi}(2a)E_{1,k-1}(2a/d) + \sum_{a=m+1}^{q}\overline{\chi}(2a)E_{1,k-1}(2a/d).$$

For $m < a \le q$ put $c = q-a$. Then $2a = d-2c-1$, and so $\chi(2a) = \chi(-2c-1)$ and $E_{1,k-1}(2a/d) = (-1)^{k-1}E_{1,k-1}((2c+1)/d)$ by (4.3f). Thus the last sum $\sum_{a=m+1}^{q}$ can be written $-\sum_{c=0}^{n}\overline{\chi}(2c{+}1)E_{1,k-1}((2c{+}1)/d)$. Therefore we obtain (4.35). If $k=1$, we recall that $E_{1,0}(t)=1$. We consider (4.32) and $\overline{\chi}(2)$ times (4.35). Taking the sum and difference of these two equalities, we obtain (4.36).

Finally, taking $d = 3$ and 4 in (4.32) and (4.33), we obtain (4.37) and (4.38), as $E_{1,n} = 2^n E_{1,n}(1/2)$. This completes the proof.

We note here an easy fact. Taking $c = 1$ and $t = 0$ in (4.5), we obtain

$$(4.39) \qquad 4(2^k - 1)(k-1)!(2\pi i)^{-k}\zeta(k) = E_{1,\,k-1}(0) \qquad (0 < k \in 2\mathbf{Z}).$$

Since $P_k(-1) = (-2)^{k-1}E_{1,\,k-1}(0)$, (4.39) is essentially the same as the first equality of (4.19b). Comparing this with (4.25), we obtain

$$(4.40) \qquad 2(1-2^k)B_k = kE_{1,\,k-1}(0) = (-2)^{1-k}kP_k(-1) \qquad (0 < k \in 2\mathbf{Z}).$$

The relation between $L(2m+1,\, \mu_4)$ and the Euler number $E_{1,2m}$ as stated in (4.38) is classical and perhaps due to Euler. Otherwise, the equalities in the above theorem seem to be new, except for (4.33) and (4.37), which may possibly be known. In particular, taking χ to be real and $k = 1$ in (4.32), we obtain a well known class number formula for an imaginary quadratic field, which we will state in (5.9) below. Thus (4.32) includes such a classical result as a special case, but apparently it has never been stated in that general form even when $k = 1$. Notice that for $d = 8$ or 12, the right-hand side of (4.34) contains only one nonvanishing term, and so we can state formulas similar to (4.37) and (4.38) in such cases. We will not give their explicit forms here, since they are included in (6.3) and (6.5) below as special cases. It should also be noted that $E_{c,\,k-1}(t)$ is a polynomial in t of degree $k-1$, while $B_k(t)$ is of degree k.

Since $E_{1,0}(t) = 1$, the formulas of Theorem 4.14 involving $E_{1,k-1}$ have simple forms if $k = 1$. Another simple formula is (4.27). In general, $E_{1,k-1}(t)$ and $B_k(t)$ have more terms, and the matter is not so simple. However, a similar simplification is feasible at least to the extent described in the following theorem, which gives recurrence formulas for $L(k,\, \chi)$ modified by elementary factors.

Theorem 4.15. *Let χ be a nontrivial primitive Dirichlet character of conductor d and k a positive integer such that $\chi(-1) = (-1)^k$; let $q = [(d-1)/2]$ and $m = [(k-1)/2]$.*

(i) *Define $A(k,\, \chi)$ by*

$$(4.41) \qquad A(k,\, \chi) = 2 \cdot k!(2\pi i)^{-k}G(\overline{\chi})L(k,\, \chi).$$

Then

$$(4.42) \qquad A(k,\, \chi) = \frac{-1}{d^k}\sum_{a=1}^{d-1}\overline{\chi}(a)a^k - \frac{1}{k+1}\sum_{\nu=1}^{m}\binom{k+1}{2\nu+1}A(k-2\nu,\, \chi).$$

(ii) *Suppose d is odd or $d > 4$. Define $\Lambda(k,\, \chi)$ by*

$$(4.43) \qquad \Lambda(k,\, \chi) = \begin{cases} 4\{\overline{\chi}(2) - 2^{-k}\}(k-1)!(\pi i)^{-k}G(\overline{\chi})L(k,\, \chi) & \text{if } d \notin 2\mathbf{Z}, \\ 2(k-1)!(\pi i)^{-k}G(\overline{\chi})L(k,\, \chi) & \text{if } d \in 2\mathbf{Z}. \end{cases}$$

Then

$$\Lambda(k,\chi) = -\frac{1}{2}\sum_{\nu=1}^{m}\binom{k-1}{2\nu}\Lambda(k-2\nu,\chi) + \begin{cases} \dfrac{1}{d^{k-1}}\displaystyle\sum_{a=1}^{d-1}(-1)^a\overline{\chi}(a)a^{k-1} & \text{if } d \notin 2\mathbf{Z}, \\ \dfrac{2^{k-1}}{d^{k-1}}\displaystyle\sum_{a=1}^{q}\overline{\chi}(a)a^{k-1} & \text{if } d \in 2\mathbf{Z}. \end{cases} \tag{4.44}$$

PROOF. By (4.26), $A(k,\chi) = -\sum_{a=1}^{d-1}\overline{\chi}(a)B_k(a/d)$, and by (4.8b) we have

$$(k+1)B_k(t) = (k+1)t^k - \sum_{\mu=0}^{k-1}\binom{k+1}{\mu}B_\mu(t).$$

By (4.28), $\sum_{a=1}^{d-1}\overline{\chi}(a)B_\mu(a/d) = 0$ if $\mu - k \notin 2\mathbf{Z}$. The sum is 0 also for $\mu = 0$, as $B_0(t) = 1$. Thus we obtain (4.42). Similarly we have, by (4.3b),

$$E_{1,k-1}(t) = t^{k-1} - \frac{1}{2}\sum_{\lambda=1}^{k-1}\binom{k-1}{\lambda-1}E_{1,\lambda-1}(t).$$

Therefore we obtain (4.44) by combining (4.34) and (4.35) with (4.30) and (4.29).

If $k = 2$, we can state clear-cut formulas as follows.

Corollary 4.16. *Let χ be a nontrivial Dirichlet character of conductor d such that $\chi(-1) = 1$ and let $q = [(d-1)/2]$. Then*

$$\pi^{-2}G(\overline{\chi})L(2,\chi) = \frac{4}{d(\chi(2)-4)}\sum_{a=1}^{q}\overline{\chi}(a)a, \tag{4.45}$$

$$\pi^{-2}G(\overline{\chi})L(2,\chi) = \frac{1}{d^2}\sum_{a=1}^{d-1}\overline{\chi}(a)a^2. \tag{4.46}$$

PROOF. Take $k = 2$ in (4.32). Since $E_{1,1}(t) = t - 1/2$, the sum of (4.32) becomes $2\sum_{a=1}^{q}\overline{\chi}(a)a/d - \sum_{a=1}^{q}\overline{\chi}(a)/2$, and so we obtain (4.45) in view of (4.31a). Similarly take $k = 2$ in (4.26). Since $B_2(t) = t^2 - t + 1/6$, we obtain (4.46) in view of (4.31b). We can also take $k = 2$ in (4.44), but obtain nothing better than (4.45).

4.17. We end this section by mentioning a classical formula and its analogues. First, as an immediate consequence of (4.8d) we obtain

$$B_{n+1}(m+1) - B_{n+1} = (n+1)\sum_{k=0}^{m}k^n \qquad (0 \le n \in \mathbf{Z}), \tag{4.47}$$

which is often cited in connection with the Euler-Maclaurin formula. (We understand that $0^0 = 1$.) Likewise, from (4.3d) we can easily derive

$$(4.48)\qquad E_{c,n}(0) - \mathbf{e}(-m\alpha)E_{c,n}(m) = \{1 - \mathbf{e}(-\alpha)\} \sum_{k=0}^{m-1} \mathbf{e}(-k\alpha)k^n,$$

where $c = -\mathbf{e}(\alpha)$, $\alpha \in \mathbf{R}$, $\notin \mathbf{Z}$. In particular, for $\alpha = 1/2$ we have

$$(4.49)\qquad E_{1,n}(0) - (-1)^m E_{1,n}(m) = 2 \sum_{k=0}^{m-1} (-1)^k k^n$$

with the classical Euler polynomial $E_{1,n}(t)$. Apparently no previous researchers took the trouble of stating this rather nice-looking formula. Notice that (4.48) and (4.49) are of degree n in the parameter m, whereas (4.47) is of degree $n+1$.

We will present some more formulas for $L(k, \chi)$ in Section 6, but before doing so, we first consider applications of those we already have to certain class number formulas.

5. The class number of a cyclotomic field

5.1. The value $L(1, \chi)$ is closely related to the class number of a cyclotomic field, an imaginary quadratic field in particular. Let us now discuss this topic in the easiest cases and present some new class number formulas, assuming that the reader is familiar with some basic facts on cyclotomic fields. For an algebraic number field M of finite degree we denote by $\mathfrak{r}_M$, D_M, R_M, ζ_M, h_M, and w_M the maximal order, discriminant, regulator, Dedekind zeta function, class number of M and the number of roots of unity in M. It is a well known result of Dedekind that

$$(5.1)\qquad \lim_{s\to 1}(s-1)\zeta_M(s) = \frac{2^{r_1}(2\pi)^{r_2} R_M h_M}{w_M |D_M|^{1/2}},$$

where r_1 resp. r_2 is the number of real resp. imaginary archimedean primes of M. We apply this to a subfield of $\mathbf{Q}_{\mathrm{ab}}$. Any subfield M of $\mathbf{Q}_{\mathrm{ab}}$ is either totally real or totally imaginary. In the former case, $r_1 = [M : \mathbf{Q}]$ and $r_2 = 0$; in the latter case $r_1 = 0$ and $r_2 = [M : \mathbf{Q}]/2$.

We now fix a totally imaginary finite extension K of $\mathbf{Q}$ contained in $\mathbf{Q}_{\mathrm{ab}}$ and put $F = \{x \in K \mid x^\rho = x\}$, where ρ is the restriction of complex conjugation to K. Then F is totally real, $w_F = 2$, and $[K : F] = 2$. Let $[K : \mathbf{Q}] = 2t$. From (5.1) we obtain

$$(5.2)\qquad (\zeta_K/\zeta_F)(1) = \frac{2\pi^t}{w_K} \cdot \frac{h_K}{h_F} \cdot \frac{R_K}{R_F} \cdot \left|\frac{D_F}{D_K}\right|^{1/2}.$$

Suppose $K \subset \mathbf{Q}(\zeta)$ with $\zeta = \mathbf{e}(1/m)$, where m is a positive integer that is either odd or divisible by 4. Then $\mathrm{Gal}(\mathbf{Q}(\zeta)/\mathbf{Q})$ is isomorphic to $(\mathbf{Z}/m\mathbf{Z})^\times$, and $\mathrm{Gal}(\mathbf{Q}(\zeta)/K)$ to a subgroup H of $(\mathbf{Z}/m\mathbf{Z})^\times$. We can show that

$$(5.3)\qquad \zeta_K(s) = \zeta_F(s) \prod_{\chi \in X} L(s, \chi),$$

where X is the set of all primitive Dirichlet characters χ such that $\chi(-1) = -1$, the conductor of χ divides m, and $\chi(H) = 1$. Thus we obtain the following formula:

$$\frac{h_K}{h_F} = \frac{w_K}{2\pi^t} \cdot \frac{R_F}{R_K} \cdot \left|\frac{D_K}{D_F}\right|^{1/2} \prod_{\chi \in X} L(1, \chi). \tag{5.4}$$

In the simplest case we take K to be an imaginary quadratic field of discriminant $-d$. Then $\zeta_K(s) = L(s, \chi)\zeta(s)$, where χ is a unique real Dirichlet character of conductor d such that $\chi(-1) = -1$, that is, $\chi(a) = \left(\frac{-d}{a}\right)$, and (5.4) takes the form

$$h_K = \frac{w_K\sqrt{d}}{2\pi} L(1, \chi). \tag{5.5}$$

This formula was first proved by Dirichlet in a somewhat different context. The functional equations of $\zeta(s)$ and $\zeta_K(s)$ show that $R(s, \chi) = R(1 - s, \chi)$, which together with (4.24) proves that

$$G(\chi) = i\sqrt{d} \text{ if } \chi(a) = \left(\frac{-d}{a}\right). \tag{5.6}$$

This argument, due to Hecke, is applicable also to a quadratic character χ such that $\chi(-1) = 1$, and even to any Hecke character of a number field corresponding to a quadratic extension; see [S97, (A6.3.4), (A6.4.1)]. Combining (4.27), (5.5), and (5.6), we obtain a classical formula

$$h_K = -\frac{w_K}{2d} \sum_{a=1}^{d} \chi(a)a. \tag{5.7}$$

5.2. Let χ be a primitive (not necessarily real) Dirichlet character of conductor d such that $\chi(-1) = -1$. Since $E_{1,0}(t) = 1$, from (4.32) with $k = 1$ we obtain

$$G(\overline{\chi})L(1, \chi) = \frac{\pi i}{2 - \chi(2)} \sum_{a=1}^{q} \overline{\chi}(a), \quad q = [(d - 1)/2]. \tag{5.8}$$

Comparing this with (4.27), we obtain

$$\sum_{a=1}^{d} \chi(a)a = \frac{d}{\overline{\chi}(2) - 2} \sum_{a=1}^{q} \chi(a). \tag{5.8a}$$

In particular, if χ is real and $K = \mathbf{Q}(\sqrt{-d})$, then (5.5) combined with (5.6) and (5.8) shows that

$$h_K = \frac{w_K}{2(2 - \chi(2))} \sum_{a=1}^{q} \chi(a), \quad q = [(d - 1)/2], \tag{5.9}$$

where the symbols are the same as in (5.5) and (5.7). This formula is also classical. Thus (4.32) or its special case (5.8) can be viewed as a generalization of (5.9). We will present several new formulas for h_K in Corollary 6.4 below.

5.3. We now consider the field $K = \mathbf{Q}(\zeta)$ with $\zeta = \mathbf{e}(1/m)$, $0 < m \in \mathbf{Z}$. We will soon specialize this to the case where m is a prime power, but first recall some basic facts on K under the assumption that m is either odd or divisible by 4. Let $F = \{x \in K \,|\, x^\rho = x\}$ as in §5.1. We put

(5.10) $$t = [K : \mathbf{Q}]/2.$$

Here are some basic facts that can be found in most textbooks on algebraic number theory:

(5.11a) $w_K = 2m$ *if* m *is odd, and* $w_K = m$ *if* $4|m$.

(5.11b) *A prime number* p *is ramified in* K *if and only if* $p|m$.

(5.11c) *For a prime number* p *not dividing* m, *let* f *be the smallest positive integer such that* $p^f - 1 \in m\mathbf{Z}$ *and let* $g = 2t/f$. *Then* p *splits into exactly* g *prime ideals in* K *each of which has degree* f.

(5.11d) $\mathfrak{r}_K = \mathbf{Z}[\zeta]$.

(5.11e) $h_K/h_F \in \mathbf{Z}$.

To prove the last statement, take the maximal unramified abelian extension J of F. Since K is ramified over F at every archimedean prime, we see that $K \not\subset J$. Thus $h_F = [J : F] = [JK : K]$, which divides h_K, as JK is an unramified abelian extension of K. This proves (5.11e).

We now assume that $m = \ell^r$ with a prime ℓ and $0 < r \in \mathbf{Z}$; $r > 1$ if $\ell = 2$. Then $2t = \ell^{r-1}(\ell - 1)$ and the following statements hold:

(5.12a) $(1-\zeta)\mathfrak{r}_K$ *is a prime ideal in* K *and* $\ell\mathfrak{r}_K = (1-\zeta)^{2t}\mathfrak{r}_K$.

(5.12b) $|D_K| = \ell^e$, *where* $e = r\ell^r - (r+1)\ell^{r-1})$ *if* $\ell \neq 2$, *and* $e = (r-1)2^{r-1}$ *if* $\ell = 2$; $|D_F| = \ell^{(e-1)/2}$ *if* $\ell \neq 2$, *and* $|D_F| = 2^{(e-2)/2}$ *if* $\ell = 2$.

Lemma 5.4. *If* m *is a prime power, then* $\mathfrak{r}_K^\times = W\mathfrak{r}_F^\times$, *where* W *is the set of all roots of unity in* K. *Consequently,* $R_K = 2^{t-1}R_F$.

PROOF. For $\alpha \in \mathfrak{r}_K^\times$ let $\beta = \alpha^\rho/\alpha$. Then $|\beta^\sigma| = 1$ for every $\sigma \in \mathrm{Gal}(K/\mathbf{Q})$. By Kronecker's theorem, $\beta \in W$. Suppose m is odd. Then $\beta = \varepsilon\gamma^2$ with $\gamma \in W$ and $\varepsilon = \pm 1$. If $\varepsilon = -1$, we have $(\alpha\gamma)^\rho = -\alpha\gamma$, and so $2\alpha\gamma = \alpha\gamma - (\alpha\gamma)^\rho \in d(K/F)$, where $d(K/F)$ is the different of K relative to F. This is a contradiction, since $d(K/F)$ is nontrivial (as can be seen from (5.12a)) and prime to 2. Thus $\varepsilon = 1$, and so $(\alpha\gamma)^\rho = \alpha\gamma$, which shows that $\alpha\gamma \in F$. Therefore $\alpha \in W\mathfrak{r}_F$ as expected. Suppose $m = 2^r$; let $n = 2^{r-1}$ and $\beta = \zeta^{-a}$ with $a \in \mathbf{Z}$. Then $\zeta^n = -1$. If a is even, we can put $\zeta = \gamma^2$ with $\gamma \in W$, which together with the above argument leads to the desired conclusion; so assume a to be odd. Since $\mathfrak{r}_K = \mathbf{Z}[\zeta]$, we have $\alpha = \sum_{s=0}^{n-1} c_s\zeta^s$ with $c_s \in \mathbf{Z}$. Then $\sum c_s\zeta^s = \sum c_s\zeta^{a-s}$. Hence $c_s = \pm c_{s'}$ if $s + s' \equiv a \pmod{n}$. Now the elements ζ^s and ζ^{a-s} for positive odd $s < n$ form a basis of K over $\mathbf{Q}$. In other words, $\zeta^{s'}$ with even s' can be replaced by ζ^{a-s} with some odd $s < n$. Therefore $\alpha = \sum_{s\text{ odd}} c_s\left(\zeta^s \pm \zeta^{a-s}\right)$. Now $\zeta^s \pm \zeta^{a-s} = \zeta^s(1-\zeta^q)$ with some q, and this is divisible by $1-\zeta$, which is a contradiction, since α is a unit. Once we know that $\mathfrak{r}_K^\times = W\mathfrak{r}_F^\times$, the assertion concerning R_K and R_F easily follows from the definition of the regulator.

Lemma 5.5. *Let $m = \ell^r$ with a prime number ℓ and $0 < r \in \mathbf{Z}$; assume $r > 3$ if $\ell = 2$. Let X be the set of all primitive Dirichlet characters χ such that $\chi(-1) = -1$ and the conductor of χ divides m. Then*

$$\ell \neq 2: \quad \prod_{\chi\in X} G(\chi) = i^t\ell^{(e+1)/4}, \quad e = r\ell^r - (r+1)\ell^{r-1},$$

$$\ell = 2: \quad \prod_{\chi\in X} G(\chi) = \sqrt{2}\cdot 2^c, \quad c = (r-1)2^{r-3}.$$

Proof. We can put $X = \bigsqcup_{s=1}^r X_s$, where X_s is the set of all $\chi \in X$ with conductor ℓ^s. Put $e_s = \#(X_s)$.

(I) We first consider the case of odd ℓ. We easily see that $e_1 = (\ell-1)/2$ and $e_s = \ell^{s-2}(\ell-1)^2/2$ if $s > 1$. For $\chi \in X$ we have $\chi = \overline{\chi}$ only if $\ell + 1 \in 4\mathbf{Z}$, in which case there is exactly one such χ, which belongs to X_1. Thus $\chi \neq \overline{\chi}$ for $\chi \in X_s$ if $s > 1$ or $\ell - 1 \in 4\mathbf{Z}$. Therefore by (4.22) we have

$$\prod_{\chi\in X_s} G(\chi) = (-\ell^s)^{e_s/2} \quad \text{if} \quad s > 1 \quad \text{or} \quad \ell - 1 \in 4\mathbf{Z}. \tag{5.13}$$

If $\ell + 1 \in 4\mathbf{Z}$, we have, by (5.6), $\prod_{\chi\in X_1} G(\chi) = i\ell^{1/2}(-\ell)^{(\ell-3)/4}$, which can be written

$$\prod_{\chi\in X_1} G(\chi) = (i\ell^{1/2})^{(\ell-1)/2}. \tag{5.14}$$

This is true also when $\ell - 1 \in 4\mathbf{Z}$. To simplify our notation, put $q = (\ell-1)/2$. Then from (5.13) and (5.14) we obtain $\prod_{\chi\in X} G(\chi) = i^a\ell^{b/2}$ with

$$a = q + \sum_{s=2}^r e_s \quad \text{and} \quad b = q + \sum_{s=2}^r se_s.$$

We easily find that $a = q\ell^{r-1} = t$. To calculate b, we note an elementary equality

$$\sum_{n=1}^k nx^n = \frac{kx^{k+2} - (k+1)x^{k+1} + x}{(x-1)^2}, \tag{5.15}$$

where x is an indeterminate. Now $\sum_{s=2}^r s\ell^{s-2} = 2\sum_{s=2}^r \ell^{s-2} + \sum_{s=2}^r (s-2)\ell^{s-2}$. Applying (5.15) to the last sum, we eventually find that $2b = r\ell^r - (r+1)\ell^{r-1} + 1$.

(II) Next we take $\ell = 2$. Then $e_1 = 0$, $e_2 = 1$, and $e_s = 2^{s-3}$ if $s > 2$. The set X_2 resp. X_3 consists of a real character of conductor 4 resp. 8; otherwise $\chi \neq \overline{\chi}$. Then by the same type of reasoning as for odd ℓ, we find the result as stated in our lemma.

Lemma 5.6. *Let $m = \ell^r$ as in Lemma 5.5, and let p be a prime number other than ℓ. Let Z be the subgroup of $(\mathbf{Z}/m\mathbf{Z})^\times$ generated by p and let $f = [Z:1]$. (In other words, f is the smallest positive integer such that $p^f - 1 \in m\mathbf{Z}$.) Then*

$$\prod_{\chi \in X} [1 - \chi(p)p^{-s}] = \begin{cases} (1 + p^{-fs/2})^{2t/f} & \text{if } \ -1 \in Z, \\ (1 - p^{-fs})^{t/f} & \text{if } \ -1 \notin Z. \end{cases} \tag{5.16}$$

Moreover, if ℓ is odd, then f is even if and only if $-1 \in Z$.

PROOF. The canonical isomorphism of $(\mathbf{Z}/m\mathbf{Z})^\times$ onto $\mathrm{Gal}(K/\mathbf{Q})$ sends -1 to ρ and Z to the decomposition group of p. Therefore the Euler p-factor of ζ_K is $(1 - p^{-fs})^{2t/f}$. Let $\mathfrak{p}$ be the prime factor of p in K. Then $\mathfrak{p}^\rho = \mathfrak{p}$ if and only if $-1 \in Z$. Thus the Euler p-factor of ζ_F equals $(1 - p^{-fs/2})^{2t/f}$ if $-1 \in Z$, and $(1 - p^{-fs})^{t/f}$ if $-1 \notin Z$. Taking the quotient ζ_K/ζ_F, we obtain (5.16). The last assertion follows from the fact that $(\mathbf{Z}/m\mathbf{Z})^\times$ is cyclic if ℓ is odd.

We now present formulas for h_k/h_F different from classical ones.

Theorem 5.7. *Let K and F be as in §5.3 with $m = \ell^r$, $0 < r \in \mathbf{Z}$, where ℓ is a prime number and $r > 2$ if $\ell = 2$. Let X be the set of all primitive Dirichlet characters χ such that $\chi(-1) = -1$ and the conductor of χ divides m. Then the following assertions hold:*

(i) *Suppose $\ell \neq 2$; let f be the smallest positive integer such that $2^f - 1 \in m\mathbf{Z}$. Put $A = (2^{f/2} + 1)^{2t/f}$ or $(2^f - 1)^{t/f}$ according as f is even or odd. For each $\chi \in X$ of conductor ℓ^s put $q_\chi = (\ell^s - 1)/2$. Then we have*

$$\frac{h_K}{h_F} = 2^{1-t}\ell^r A^{-1} \prod_{\chi \in X} \left\{ \sum_{a=1}^{q_\chi} \chi(a) \right\}. \tag{5.17}$$

Moreover, the sum $\sum_{a=1}^{q_\chi} \chi(a)$ can be replaced by

$$\frac{2}{1 + \overline{\chi}(2)} \sum_{a=1}^{k_\chi} \chi(a) \quad \text{with} \quad k_\chi = [q_\chi/2] \quad \text{if} \quad \chi(2) \neq -1, \tag{5.17a}$$

$$\frac{2}{1 - \overline{\chi}(2)} \sum_{c=0}^{n_\chi} \chi(2c+1) \quad \text{with} \quad n_\chi = [(q_\chi - 1)/2] \quad \text{if} \quad \chi(2) \neq 1. \tag{5.17b}$$

(ii) *Suppose $\ell = 2$; let Y be the set of all $\chi \in X$ of conductor > 4. For $\chi \in Y$ of conductor 2^s, put $b_\chi = 2^{s-2}$. Then*

(5.18) $$\frac{h_K}{h_F} = 2^\gamma \prod_{\chi\in Y}\left\{\sum_{a=1}^{b_\chi-1}\chi(a)\right\} \quad \textit{with} \quad \gamma = r-1-2^{r-2}.$$

PROOF. We employ (5.4). We have $w_K = 2m$ if $\ell \neq 2$ and $w_K = m$ if $\ell = 2$; $R_F/R_K = 2^{1-t}$ as shown in Lemma 5.4; $|D_K/D_F|$ is given by (5.12b). We have to determine $\prod_{\chi\in X} L(1,\chi)$. Suppose $\ell \neq 2$. Then we use (5.8) which involves $2-\chi(2)$ and the Gauss sum. As for the latter, we employ Lemma 5.5. As for the former, from Lemma 5.6 we obtain

(5.19) $$\prod_{\chi\in X}[2-\chi(2)] = \begin{cases} (2^{f/2}+1)^{2t/f} & \text{if } f\in 2\mathbf{Z},\\ (2^f-1)^{t/f} & \text{if } f\notin 2\mathbf{Z}.\end{cases}$$

Combining all these factors, we obtain (5.17). The sum $\sum_{a=1}^{q_\chi}\chi(a)$ can be replaced by (5.17a, b) by virtue of (4.36). If $\ell = 2$, we use (4.34) instead of (5.8). Special care must be given to the character of conductor 4, which is included in X but not in Y. Observing that c of Lemma 5.5 equals $e/4$ with e of (5.12b), we obtain (5.18) for $r > 3$. Actually we see that (5.18) is true even for $r = 2$ and 3.

Remark. For odd ℓ, it often happens that 2 is a primitive root modulo m, in which case we have $f = 2t$, and so $A = 2^t+1$. Thus we can put $A = \ell^r I$ with a positive integer I, and the factor $\ell^r A^{-1}$ in (5.17) equals I^{-1}.

5.8. Let us now state the classical formulas for h_K/h_F that can be found in the standard literature on this topic:

(5.20) $$\frac{h_K}{h_F} = B\prod_{\chi\in X}\left\{-\sum_{a=1}^{c_\chi}\chi(a)a\right\} \quad \text{with}$$

$$B = \begin{cases} 2^{1-t}\ell^\alpha, & \alpha = r - 2^{-1}(r\ell^r - (r+1)\ell^{r-1}+1) \ \text{ if } \ \ell\neq 2,\\ 2^\beta, & \beta = r-1-2^{r-2}r \ \text{ if } \ \ell = 2.\end{cases}$$

Here c_χ is the conductor of χ. This can be obtained from (5.4) by applying (4.27) to $L(1,\chi)$ and employing Lemmas 5.4 and 5.5.

Clearly formulas (5.17) and (5.18) are of "smaller sizes" than (5.20). That is especially so for (5.18). As for (5.17), we derived it from (5.20) combined with (5.8a). Equality (5.8a) is an old well known fact at least for real χ. Even for non-real χ, it must have been known at least to some experts, but apparently nobody tried to state (5.17).

In stating formulas for h_K/h_F, we confined ourselves to the case where $K = \mathbf{Q}(\zeta)$, $\zeta = \mathbf{e}(1/m)$, with a prime power m. A formula of type (5.20) for more general cyclotomic fields is known; see [Ha]. We can of course state analogues of (5.17) and (5.18) for such fields, whose precise statements may be left to the reader.

6. Some more formulas for $L(k, \chi)$

6.1. Let us now state several sum expressions for $L(k, \chi)$ different from those of Section 4. The first type has fewer terms than the sum of (4.32), and are technically more complex than the second type, which follows from (4.5) rather easily. We present the first type as eight formulas depending on the nature of the Dirichlet character and the parameter c of the generalized Euler polynomial $E_{c,n}$. One formula, (6.2), is somewhat different from the other seven. Though it may be possible to state those seven as a single formula, we do not do so, as such would make it cumbersome and less easy to understand. In the following theorem, the product $\chi\lambda$ for two characters χ and λ means the character defined by $(\chi\lambda)(m) = \chi(m)\lambda(m)$, which can be imprimitive if the conductors of χ and λ are not relatively prime. We first prove

Lemma 6.2. *Let ε denote ± 1, and given two positive integers d and s, let $g = [d/s]+1$ or $g = -[d/s]$ according as $\varepsilon = 1$ or -1. Then $0 \le (\varepsilon/s)+(j/d) \le 1$ for every $j \in \mathbf{Z}$ such that $1-g \le j \le d-g$. Moreover, $0 < (\varepsilon/s)+(j/d) < 1$ for all such j's if $d/s \notin \mathbf{Z}$.*

PROOF. If $g = [d/s]+1$, then $g-1 \le d/s < g$, and so $(g-1)/d \le 1/s < g/d$. For $1-g \le j \le d-g$ we have $(1-g)/d \le j/d \le 1-g/d$, and therefore $0 \le (1/s)+(j/d) < 1$. If $g = -[d/s]$, then $-g \le d/s < 1-g$, and so $(g-1)/d < -1/s \le g/d$, and we obtain $0 < (-1/s)+(j/d) \le 1$. If $d/s \notin \mathbf{Z}$, we easily see that $0 < (\varepsilon/s)+(j/d) < 1$ in both cases.

Theorem 6.3. *Let χ and λ be primitive Dirichlet characters and μ_3, μ_4 be as in Theorem 4.14 (iv); let d be the conductor of χ and k a positive integer; further let ε denote ± 1. Then the following assertions hold:*

(i) *Suppose d is odd, > 3, and $\chi(-1) = (-1)^{k+1}$; suppose also $k > 1$ if $3|d$; let $g = [d/3]+1$. Then*

$$\sum_{j=1-g}^{d-g} (-1)^j\overline{\chi}(j)E_{1,k-1}\left(\tfrac{1}{3}+\tfrac{j}{d}\right) = (k-1)!2\sqrt{-3}(\pi i)^{-k}\overline{\chi}(2)G(\overline{\chi})\{1+\chi(2)2^{-k}\}L(k, \chi\mu_3). \tag{6.1}$$

(ii) *Suppose $d = 2m+1$ with $0 < m \in \mathbf{Z}$ and $\chi(-1) = (-1)^{k+1}$. Then*

$$\sum_{j=1}^{m}(-1)^j\overline{\chi}(j)E_{1,k-1}\left(\tfrac{1}{2}+\tfrac{j}{d}\right) = (k-1)!2i(\pi i)^{-k}\overline{\chi}(2)G(\overline{\chi})L(k, \chi\mu_4). \tag{6.2}$$

(iii) *Suppose d is odd, $\lambda(m) = \left(\frac{\delta}{m}\right)$ with $\delta = \pm 8$, and $(\chi\lambda)(-1) = (-1)^k$. Then*

$$\sum_{j=1-g}^{d-g} (-1)^j\overline{\chi}(j)E_{1,k-1}\left(\tfrac{1}{4}+\tfrac{j}{d}\right) = (k-1)!\sqrt{\delta}(\pi i)^{-k}\overline{\chi}(2)G(\overline{\chi})L(k, \chi\lambda), \tag{6.3}$$

where $g = [d/4]+1$. *The formula is valid even for* $d = 1$ *and trivial* χ, *in which case the sum on the left-hand side is* $E_{1,k-1}(1/4)$ *and* $G(\overline{\chi}) = 1$.

(iv) *For* d, λ, χ, *and* k *of* (iii) *we have also*

$$(6.4)\qquad \sum_{j=1-g}^{d-g} \mathbf{e}(-dj/4)\overline{\chi}(j)E_{-i,k-1}\left(\tfrac{1}{2}+\tfrac{j}{d}\right) = (k-1)!\sqrt{-2}\cdot 2^k(\pi i)^{-k}\overline{\chi}(4)G(\overline{\chi})L(k,\,\chi\lambda),$$

where $g = [d/2]+1$. *The formula is valid even for* $d = 1$ *and trivial* χ, *in which case the sum on the left-hand side is* $E_{-i,k-1}(1/2)$ *and* $G(\overline{\chi}) = 1$.

(v) *Suppose* d *is odd,* > 3, $\lambda(m) = \left(\frac{3}{m}\right)$, *and* $\chi(-1) = (-1)^k$. *Then*

$$(6.5)\qquad \sum_{j=1-g}^{d-g} (-1)^j\overline{\chi}(j)E_{1,k-1}\left(\tfrac{1}{6}+\tfrac{j}{d}\right) = (k-1)!2\sqrt{3}(\pi i)^{-k}\overline{\chi}(2)G(\overline{\chi})L(k,\,\chi\lambda),$$

where $g = [d/6]+1$. *The formula is valid even for* $d = 1$ *and trivial* χ, *in which case* $G(\overline{\chi}) = 1$ *and the sum on the left-hand side is* $E_{1,k-1}(1/6)$.

(vi) *For* d *prime to 6 and* λ, χ, k *of* (v) *we have*

$$(6.6)\qquad \sum_{j=1-g}^{d-g} \mathbf{e}(-dj/6)\overline{\chi}(j)E_{c,k-1}\left(\tfrac{1}{2}+\tfrac{j}{d}\right) = (k-1)!\cdot 3^k i(\pi i)^{-k}\overline{\chi}(6)G(\overline{\chi})L(k,\,\chi\lambda),$$

where $g = [d/2]+1$ *and* $c = \mathbf{e}(2/3)$. *The formula is valid even for* $d = 1$ *and trivial* χ, *in which case* $G(\overline{\chi}) = 1$ *and the sum on the left-hand side is* $E_{c,k-1}(1/2)$.

(vii) *Suppose* d *is prime to 3; let* λ *be the Dirichlet character modulo 9 such that* $\lambda(2) = \mathbf{e}(2/3)$. *Then*

$$(6.7)\qquad \sum_{j=1-g}^{d-g} \mathbf{e}(-dj/3)\overline{\chi}(j)E_{c,k-1}\left(\tfrac{\varepsilon}{3}+\tfrac{j}{d}\right) = (k-1)!\mathbf{e}(\varepsilon/9)\{1+\mathbf{e}(1/6)\}$$
$$\cdot(2\pi i)^{-k}3^k\overline{\chi}(3)G(\overline{\chi})\cdot\begin{cases} L(k,\,\chi\lambda^\varepsilon) & \text{if } \ \chi(-1) = (-1)^k, \\ L(k,\,\mu_3\chi\lambda^\varepsilon) & \text{if } \ \chi(-1) = (-1)^{k+1}, \end{cases}$$

where $c = -\mathbf{e}(1/3)$, *and* $g = [d/3]+1$ *or* $g = -[d/3]$ *according as* $\varepsilon = 1$ *or* $\varepsilon = -1$. *The formula is valid even for* $d = 1$ *and trivial* χ, *in which case* $G(\overline{\chi}) = 1$ *and the sum on the left-hand side is* $E_{c,k-1}(1/3)$ *if* $\varepsilon = 1$ *and* $E_{c,k-1}(2/3)$ *if* $\varepsilon = -1$.

(viii) *Suppose* d *is odd; let* λ *be the Dirichlet character modulo 16 such that* $\lambda(5) = i$ *and* $(\chi\lambda)(-1) = (-1)^k$. *Then*

$$(6.8)\qquad \sum_{j=1-g}^{d-g} \mathbf{e}(-dj/4)\overline{\chi}(j)E_{-i,k-1}\left(\tfrac{\varepsilon}{4}+\tfrac{j}{d}\right) = (k-1)!(1+i)\mathbf{e}\left(\tfrac{\varepsilon}{16}\right)\cdot 2^k(\pi i)^{-k}\overline{\chi}(4)G(\overline{\chi})L(k,\,\chi\lambda^\varepsilon),$$

where $g = [d/4] + 1$ or $g = -[d/4]$ according as $\varepsilon = 1$ or $\varepsilon = -1$. The formula is valid even for $d = 1$ and trivial χ, in which case $G(\overline{\chi}) = 1$ and the sum on the left-hand side is $E_{-i,k-1}(1/4)$ if $\varepsilon = 1$ and $E_{-i,k-1}(3/4)$ if $\varepsilon = -1$. (Once χ and k are given, λ is uniquely determined by the conditions $\lambda(5) = i$ and $(\chi\lambda)(-1) = (-1)^k$.)

PROOF. We first consider Case (v), which has a feature that the other cases lack. We use formula (4.5) with $\alpha = 1/2$ and $t = (1/6) + (j/d)$ with j in the sum expression of (6.5). By Lemma 6.2 we have $0 < t < 1$, and so (4.5) is applicable to the present setting for $n \geq 0$. Put $j = pd + 2q$ with $p, q \in \mathbf{Z}$. Then for $h \in \mathbf{Z}$ we have

$$\mathbf{e}\big((h+\alpha)t\big) = \mathbf{e}(1/12)\mathbf{e}(h/6)\mathbf{e}(p/2)\mathbf{e}\big((2h+1)q/d\big).$$

Clearly $\mathbf{e}(p/2) = \mathbf{e}(j/2) = (-1)^j$. Therefore, by (4.5), the left-hand side of (6.5) equals

$$2\mathbf{e}(1/12)(k-1)!(\pi i)^{-k}\sum_{h\in\mathbf{Z}}(2h+1)^{-k}\mathbf{e}(h/6)\sum_{j}\overline{\chi}(j)\mathbf{e}\big((2h+1)q/d\big).$$

Since $\chi(j) = \chi(2q)$ and q runs over $\mathbf{Z}/d\mathbf{Z}$, the sum $\sum_j$ equals $\overline{\chi}(2)G(\overline{\chi})\chi(2h+1)$. Put $\sum_{h\in\mathbf{Z}}\mathbf{e}(h/6)\chi(2h+1)(2h+1)^{-k} = \sum_{n=1}^{\infty} b_n n^{-k}$. Since $\chi(-1) = (-1)^k$, we have $b_n = \chi(n)\big\{\mathbf{e}((n-1)/12) + \mathbf{e}((-n-1)/12)\big\}$. We easily see that this equals $\{1 + \mathbf{e}(5/6)\}\chi(n)\lambda(n)$. (Notice that $b_n = 0$ if $3|n$.) Therefore we obtain (6.5). If $d = 1$ and χ is trivial, we consider $E_{1,k-1}(1/6)$ and put $G(\overline{\chi}) = 1$. Then the above argument is valid in that case too. Also, if $k = 1$, we have to invoke Lemma 4.3 for the same reason as in the proof of Theorems 4.12 and 4.14.

The other cases except Case (ii) can be proven basically in the same fashion. We take $\alpha = 1/3$ in Case (vii), $\alpha = 1/6$ in Case (vi), $\alpha = 1/4$ in Cases (iv) and (viii), and $\alpha = 1/2$ in the remaining cases. We also take $j = pd + qr$ with $p, q \in \mathbf{Z}$, where $r = \alpha^{-1}$. If $3|d$ in Case (i), then $(1/3) + (j/d) = 0$ for $j = 1 - g$, and therefore (4.5) is applicable only to the case $k > 1$.

In Case (ii), by the same technique with $\alpha = r^{-1} = 1/2$ we first obtain

$$\sum_{j=-m}^{m}(-1)^j\overline{\chi}(j)E_{1,k-1}\big(\tfrac{1}{2} + \tfrac{j}{d}\big) = (k-1)!4i(\pi i)^{-k}\overline{\chi}(2)G(\overline{\chi})L(k, \chi\mu_4), \tag{6.9}$$

where $m = [d/2]$. The left-hand side can be written

$$\sum_{j=1}^{m}(-1)^j\overline{\chi}(j)\Big\{E_{1,k-1}\big(\tfrac{1}{2} + \tfrac{j}{d}\big) + \chi(-1)E_{1,k-1}\big(\tfrac{1}{2} - \tfrac{j}{d}\big)\Big\}.$$

By (4.3f) this equals twice the left-hand side of (6.2). Dividing by 2, we obtain (6.2).

Now some class number formulas different from (5.7) and (5.9) can be obtained by taking $k = 1$ and χ to be a real character in (4.34), (4.36), and also in the first six cases of the above theorem. Here are their explicit statements.

Corollary 6.4. *Let d_0 denote a positive squarefree integer > 1 and h_K the class number of the field K given in each case below. Except in Case (i) suppose d_0 is odd and let χ_0 be the real primitive Dirichlet character of conductor d_0. Then the following assertions hold.*

(i) *Suppose $2 < d_0 + 1 \notin 4\mathbf{Z}$ and $K = \mathbf{Q}(\sqrt{-d_0}\,)$; let χ be the primitive quadratic Dirichlet character of conductor $4d_0$ that corresponds to K, and $\nu = [d_0/2]$. Then*

$$h_K = \sum_{a=1}^{d_0-1} \chi(a) = \sum_{b=1}^{\nu} \chi(2b-1). \tag{6.10}$$

(ii) *Suppose $d_0 = 4\mu + 1$ with $0 < \mu \in \mathbf{Z}$ and let $K = \mathbf{Q}(\sqrt{-d_0}\,)$. Then*

$$h_K = \chi_0(2) \sum_{a=1}^{2\mu} (-1)^a \chi_0(a). \tag{6.11}$$

(iii) *Suppose $d_0 = 4\mu + 3$ with $0 < \mu \in \mathbf{Z}$ and let $K = \mathbf{Q}(\sqrt{-d_0}\,)$. Then*

$$h_K = \begin{cases} \displaystyle\sum_{a=1}^{\mu} \chi(a) & (\mu \notin 2\mathbf{Z}), \\ \displaystyle\frac{1}{3}\sum_{c=0}^{\mu} \chi(2c+1) & (\mu \in 2\mathbf{Z}). \end{cases} \tag{6.12}$$

(iv) *Let $K = \mathbf{Q}(\sqrt{-2d_0}\,)$ and $\mu = [d_0/4]$. Then*

$$h_K = 2\chi_0(2) \sum_{a=1}^{\mu} (-1)^a \chi_0(a) \quad \textit{if} \quad d_0 = 4\mu + 1, \tag{6.13}$$

$$h_K = 2\sum_{a=0}^{\mu} (-1)^a \chi_0(2a+1) = 2\chi_0(2) \sum_{b=\mu+1}^{2\mu+1} (-1)^b \chi_0(b) \ \textit{ if } \ d_0 = 4\mu + 3. \tag{6.14}$$

(v) *Suppose that d_0 is prime to 3 and $d_0 + 1 \in 4\mathbf{Z}$; let $K = \mathbf{Q}(\sqrt{-3d_0}\,)$ and $m = [d_0/2]$. Then*

$$h_K = 2\chi_0(2)\Big\{ \sum_a \chi_0(a) - \sum_b \chi_0(b) \Big\}, \tag{6.15}$$

where $1 \le a \le m$, $1 \le b \le m$, $a \equiv 1, 2 \pmod 6$, and $b \equiv 4, 5 \pmod 6$.

(vi) *Let K and d_0 be the same as in (iv)* and let $m = [d_0/6]$. Then

$$h_K = 2\chi_0(2) \sum_{a=m+1}^{(d_0-1)/2} (-1)^a \chi_0(a). \tag{6.16}$$

(vii) *Suppose that d_0 is prime to 3 and $d_0 - 1 \in 4\mathbf{Z}$; let $K = \mathbf{Q}(\sqrt{-3d_0}\,)$ and $m = [d_0/3]$. Then*

$$h_K = \frac{2}{2\chi_0(2) + 1} \sum_{a=1}^{m} (-1)^a \chi_0(a). \tag{6.17}$$

PROOF. We first recall that $E_{c,0}(t) = 1$ as noted in (4.3h). Take $k = 1$ in (4.34). Then by (5.5) we obtain (6.10), as $\chi(a) = 0$ for even a. Formula (6.12) folllows immediately from (4.36) combined with (5.5). Next we take $k = 1$ and (χ_0, d_0) to be (χ, d) in Theorem 6.3. Let us first consider (6.3). Suppose $d_0 = 4\mu + 3$. Then $\chi_0(-1) = -1$, and the left-hand side of (6.3) equals

$$\sum_{a=-\mu}^{3\mu+2} (-1)^a \chi_0(a) = \sum_{a=-\mu}^{\mu} (-1)^a \chi_0(a) + \sum_{b=\mu+1}^{3\mu+2} (-1)^b \chi_0(b)$$

$$= \sum_{a=1}^{\mu} (-1)^a \{\chi_0(a) + \chi_0(-a)\} + \sum_{b=\mu+1}^{2\mu+1} \{(-1)^b \chi_0(b) + (-1)^{d_0-b} \chi_0(d_0 - b)\}.$$

The first sum of the last line is 0; the second sum equals $2\sum_{b=\mu+1}^{2\mu+1}(-1)^b\chi_0(b)$. Putting $b = 2\mu + 1 - c$, we have $2b = d - 2c - 1$, and so the last sum equals $\chi_0(2)\sum_{c=0}^{\mu}(-1)^{c+1}\chi_0(d_0 - 2c - 1) = \chi_0(2)\sum_{c=0}^{\mu}\chi_0(2c + 1)$. Applying (6.3) to the factor $L(1, \chi\lambda)$ of (6.3), we obtain (6.14). Taking $d_0 = 4\mu + 1$ in (6.3), we similarly obtain (6.13). These formulas (6.13) and (6.14) can be obtained also from (6.4).

Formula (6.11) follows directly from (6.2) combined with (5.5) applied to the present K. To prove (6.15), we use (6.6). Notice that $d_0 \equiv 7$ or $11 \pmod{12}$. Put $\omega = \mathbf{e}(1/6)$, and suppose $d_0 - 7 \in 12\mathbf{Z}$. Then the left-hand side of (6.6) can be written $\sum_{j=-m}^{m} \chi(j)\omega^{-j}$, which equals $\sum_{j=1}^{m} \chi(j)(\omega^{-j} - \omega^{j})$. We easily see that this equals

$$-i\sqrt{3}\Big\{\sum_a \chi_0(a) - \sum_b \chi_0(b)\Big\} \tag{6.18}$$

with a and b as in (v). Applying (5.5) to $L(1, \chi_0\lambda)$, we can verify that the right-hand side of (6.6) equals $i\sqrt{3}\chi_0(6)h_K/2$. If $d_0 - 11 \in 12\mathbf{Z}$, then ω must be replaced by ω^{-1}, and we have $i\sqrt{3}$ instead of $-i\sqrt{3}$ in (6.18). Now $\chi_0(3) = \left(\frac{3}{d_0}\right)$, which equals -1 or 1 according as $d_0 \equiv 7$ or $11 \pmod{12}$. Therefore we obtain (6.15) in both cases as expected.

We derive (6.16) from (6.5). Put $m = [d_0/6]$ and $d_0 = 2q + 1$. Then the left-hand side of (6.5) becomes $\sum_{a=-m}^{d_0-m-1}(-1)^a\chi_0(a)$. We have $\sum_{a=-m}^{m}(-1)^a\chi_0(a) = 0$ as $\chi_0(-1) = -1$, and so we only have to consider $\sum_{a=m+1}^{d_0-m-1}(-1)^a\chi_0(a)$, which equals

$$\sum_{a=m+1}^{q} \Big\{(-1)^a\chi_0(a) + (-1)^{d_0-a}\chi_0(d_0 - a)\Big\} = 2\sum_{a=m+1}^{q} (-1)^a\chi_0(a).$$

Thus we obtain (6.16).

Finally, in the setting of (vii) we see that the left-hand side of (6.1) equals

$$\sum_{a=-m}^{d_0-m-1} (-1)^a\chi_0(a) = \sum_{a=-m}^{m} (-1)^a\chi_0(a) + \sum_{b=m+1}^{d_0-m-1} (-1)^b\chi_0(b).$$

The same argument as for (6.14) shows that the last sum over b vanishes, while the sum over $-m \le a \le m$ equals $2\sum_{a=1}^{m}(-1)^a\chi_0(a)$. Thus we obtain (6.17) and our proof is complete.

6.5. Though it may be possible to prove the formulas of the above corollary more directly, our point of presenting them is that they follow easily from a more general principle concerning $L(k,\,\chi)$ with $k \ge 1$. It should also be noted that they are quite different from (5.9). Take, for example, $K = \mathbf{Q}(\sqrt{-2d_0}\,)$ with $d_0 = 4\mu + 1$ as in (6.13). Then the discriminant of K is $-8d_0$, and so the sum of (5.9) has $16\mu + 3$ terms. Since $\chi(a) = 0$ for even a in this case, the number of nonvanishing terms is much smaller, but not so small as that for (6.13).

One more remark may be added, First, we can state formulas for $L(1,\,\chi)$ of various types of χ that are not necessarily real. Take χ_0 to be a primitive character whose conductor d_0 is of the form $d_0 = 4\mu+3$ and such that $\chi_0(-1) = 1$. (Such a χ_0 cannot be real.) Let $\lambda(m) = \left(\frac{-2}{m}\right)$. Then from (6.3) we can derive, by the same technique as in the proof of Corollary 6.4,

$$\pi^{-1}\sqrt{2}\,\chi_0(2)G(\chi_0)L(1,\,\overline{\chi}_0\lambda) = \sum_{a=1}^{\mu}(-1)^a\chi_0(a). \tag{6.19}$$

Similarly we can handle $L(1,\,\overline{\chi}_0\lambda)$ when $d_0 = 4\mu+1$, $\chi_0(-1) = -1$, and $\lambda(m) = \left(\frac{2}{m}\right)$, in which case a formula of type (6.14) appears. Some more formulas for $L(1,\,\chi)$ will be given in Corollary 6.7 below.

We now state some analogues of (4.32), which are not so technically involved as Theorem 6.3.

Theorem 6.6. *Let χ be a primitive character modulo d, and k a positive integer such that $\chi(-1) = (-1)^k$. Then the following assertions hold:*

(i) *Suppose d is prime to 3; let $c = -\mathbf{e}(1/3)$. Then*

$$\begin{aligned}(k-1)!(2\pi i)^{-k}G(\overline{\chi})L(k,\,\chi)&\\ = \frac{\chi(3)}{(1+c^{-1})\{3^k-\chi(3)\}}&\sum_{a=1}^{d-1}\overline{\chi}(a)\mathbf{e}(-da/3)E_{c,k-1}(a/d).\end{aligned} \tag{6.20}$$

(ii) *Suppose d is prime to 2; let $c = -\mathbf{e}(1/4)$. Then*

$$\begin{aligned}(k-1)!(2\pi i)^{-k}G(\overline{\chi})L(k,\,\chi)&\\ = \frac{\chi(4)}{(1+i)\{4^k-\chi(2)2^k\}}&\sum_{a=1}^{d-1}\overline{\chi}(a)\mathbf{e}(-da/4)E_{c,k-1}(a/d).\end{aligned} \tag{6.21}$$

(iii) *Suppose d is prime to 6; let $c = -\mathbf{e}(1/6)$. Then*

$$(k-1)!(2\pi i)^{-k}G(\overline{\chi})L(k,\,\chi) \tag{6.22}$$

$$= \frac{\chi(6)}{\mathbf{e}(1/6)\{2^k - \chi(2)\}\{3^k - \chi(3)\}} \sum_{a=1}^{d-1} \overline{\chi}(a)\mathbf{e}(-da/6)E_{c,k-1}(a/d).$$

PROOF. For simplicity, let us write F and E for $F_{c,\,k-1}$ and $E_{c,\,k-1}$. We first prove (iii). Thus $c = -\mathbf{e}(\alpha)$ with $\alpha = 1/6$. Taking $N = 6$, $q = 1$, and $t = b/d$ in (4.4c), we obtain

$$(*)\qquad F(6b/d) = 6^k \sum_{m \in 1+6\mathbf{Z}} m^{-k}\mathbf{e}(mb/d) \qquad (b \in \mathbf{Z}).$$

Clearly $F(6b/d)$ depends only on $b \pmod d$. Also, we easily see that

$$(**)\qquad \sum_{m \in 1+6\mathbf{Z}} \chi(m)m^{-k} = \{1 - \chi(2)2^{-k}\}\{1 - \chi(3)3^{-k}\}L(k, \chi).$$

If $k = 1$, the sums of $(*)$ and $(**)$ should be understood in the sense explained in a few lines below (4.4). Take integers μ and ν so that $1 = d\mu + 6\nu$. Then for $0 < a < d$ we have $a = ad\mu + 6a\nu$ and $F(a/d) = \mathbf{e}(a\mu/6)F(6a\nu/d)$ by (4.4a). Since $\mathbf{e}(a\mu/6) = \mathbf{e}(ad/6)$, we have, in view of $(*)$,

$$\begin{aligned}\sum_{a=1}^{d-1} \overline{\chi}(a)\mathbf{e}(-ad/6)F(a/d) &= \sum_{a=1}^{d-1} \overline{\chi}(a)F(6a\nu/d)\\ &= \chi(\nu)\sum_{a=1}^{d-1} \overline{\chi}(a\nu)6^k \sum_{m \in 1+6\mathbf{Z}} m^{-k}\mathbf{e}(ma\nu/d)\\ &= 6^k\chi(\nu)G(\overline{\chi}) \sum_{m \in 1+6\mathbf{Z}} \chi(m)m^{-k}.\end{aligned}$$

Applying (4.5) to $F(a/d)$ and employing $(**)$, we obtain (6.22), as $\chi(6\nu) = 1$. The other two cases with $\alpha = 1/3$ and $\alpha = 1/4$ can be proved in the same manner.

Corollary 6.7. *Let χ be a primitive character modulo d such that $\chi(-1) = -1$. Then the following assertions hold.*

(i) *If d is prime to 3, we have*

$$(6.23)\qquad (2\pi i)^{-1}G(\overline{\chi})L(1, \chi) = \frac{\chi(3)}{3 - \chi(3)}\left\{\sum_{a=1}^{[d/6]} \overline{\chi}(3a) - \sum_{b \in B} \overline{\chi}(b)\right\},$$

$$B = \bigl\{b \in \mathbf{Z} \bigm| 0 < b < d/2,\ b - d \in 3\mathbf{Z}\bigr\}.$$

(ii) *If d is prime to 2, we have*

$$(6.24)\qquad (2\pi i)^{-1}G(\overline{\chi})L(1, \chi) = \frac{\chi(4)}{2 - \chi(2)} \cdot \begin{cases} \dfrac{1}{\chi(4)+1}\displaystyle\sum_{a \in A} \overline{\chi}(a) & \text{if } \chi(4) \neq -1,\\[2ex] \dfrac{1}{\chi(4)-1}\displaystyle\sum_{b \in B} \overline{\chi}(b) & \text{if } \chi(4) \neq 1,\end{cases}$$

$$A = \bigl\{a \in \mathbf{Z} \bigm| 0 < a < d/2,\ a \equiv 0 \text{ or } -d \pmod 4\bigr\},$$
$$B = \bigl\{b \in \mathbf{Z} \bigm| 0 < b < d/2,\ b \equiv 2 \text{ or } d \pmod 4\bigr\}.$$

(iii) *If d is prime to 6, we have*

$$(6.25)\qquad (2\pi i)^{-1}G(\overline{\chi})L(1,\chi) = \frac{\chi(6)}{\{2-\chi(2)\}\{3-\chi(3)\}} \cdot\Big\{\sum_{a\in A_1}\overline{\chi}(a) + 2\sum_{a\in A_2}\overline{\chi}(a) - \sum_{b\in B_1}\overline{\chi}(b) - 2\sum_{b\in B_2}\overline{\chi}(b)\Big\},$$

$$A_1 = \{a\in\mathbf{Z}\,|\,0<a<d/2,\ a\equiv 0 \text{ or } -2d \pmod 6\},$$
$$A_2 = \{a\in\mathbf{Z}\,|\,0<a<d/2,\ a\equiv -d \pmod 6\},$$
$$B_1 = \{b\in\mathbf{Z}\,|\,0<b<d/2,\ b\equiv d \text{ or } 3 \pmod 6\},$$
$$B_2 = \{b\in\mathbf{Z}\,|\,0<b<d/2,\ b\equiv 2d \pmod 6\}.$$

PROOF. We take $k=1$ in Theorem 6.6 and use the fact that $E_{c,0}(t)=1$ as noted in (4.3h). We first consider the case $c=-\mathbf{e}(1/6)$. Put $q=(d-1)/2$ and $\zeta=\mathbf{e}(1/6)$. Then the sum on the right-hand side of (6.22) can be written

$$\sum_{a=1}^{q}\overline{\chi}(a)\{\mathbf{e}(-da/6)-\mathbf{e}((da-1)/6)\},$$

and $\mathbf{e}(-b/6)-\mathbf{e}((b-1)/6)$ is ζ, $-\zeta$, 2ζ, or -2ζ according to $b \pmod 6$. Then we obtain (6.25) with A_ν and B_ν as given there. Case (i) can be proved in the same manner. In Case (ii), by the same technique we first obtain

$$(6.26)\qquad (2\pi i)^{-1}G(\overline{\chi})L(1,\chi) = \frac{\chi(4)}{4-2\chi(2)}(X-Y),$$
$$X=\sum_{a\in A}\overline{\chi}(a),\qquad Y=\sum_{b\in B}\overline{\chi}(b)$$

with A and B as given in (6.24). On the other hand, (4.32) with $k=1$ shows that this equals $(X+Y)/(4-2\chi(2))$. Thus $X+Y=\chi(4)(X-Y)$. Therefore we can state the result as in (6.24).

If $K=\mathbf{Q}(\sqrt{-d})$ and $-d$ is the discriminant of K and $d>4$, then the formulas of Corollary 6.7 give $h_K/2$. Of course we cannot attach importance to any of such class number formulas, but we mention them simply because they follow from more general results on $L(k,\chi)$, which are well worthy of notice.

CHAPTER III

THE CASE OF IMAGINARY QUADRATIC FIELDS AND NEARLY HOLOMORPHIC MODULAR FORMS

7. Dirichlet series associated with an imaginary quadratic field

7.1. To define another type of Dirichlet series similar to $D^r_{a,N}$, we take an imaginary quadratic field K embedded in $\mathbf{C}$. We then denote by $\mathfrak{r}$ the maximal order of K and by ρ the restriction of complex conjugation to K. Fixing a $\mathbf{Z}$-lattice $\mathfrak{b}$ in K and $\alpha \in K$, put $\alpha + \mathfrak{b} = \{\xi \in K \mid \xi - \alpha \in \mathfrak{b}\}$ and

$$L^r(s;\,\alpha,\,\mathfrak{b}) = \sum_{0 \neq \xi \in \alpha+\mathfrak{b}} \xi^{-r}|\xi|^{r-2s} \qquad (0 \leq r \in \mathbf{Z}), \tag{7.1}$$

where s is a complex variable as usual. As we said in the introduction, the special values of the series of this type, when $r > 0$, are the main subject of study in this book, and will be discussed in later sections. Here we prove several basic analytic (that is, nonarithmetic) properties of (7.1) which can be obtained by the same technique as for $M^\nu_{a,N}$ by means of Fourier analysis on $\mathbf{C}/\mathfrak{b}$. Let us first recall some elementary facts. We identify $\mathbf{C}$ with $\mathbf{R}^2$ through the map $z = x + iy \mapsto (x,\,y)$. Let $\mathfrak{c}$ be a $\mathbf{Z}$-lattice in K of the form $\mathfrak{c} = \mathbf{Z}\omega_1 + \mathbf{Z}\omega_2$ with $\omega_1,\,\omega_2 \in K$. For $z = x + iy = a\omega_1 + b\omega_2$ with $a,\,b \in \mathbf{Z}$ we have

$$\begin{bmatrix} x \\ y \end{bmatrix} = 2^{-1}\begin{bmatrix} 1 & 1 \\ -i & i \end{bmatrix}\begin{bmatrix} z \\ \overline{z} \end{bmatrix} = 2^{-1}\begin{bmatrix} 1 & 1 \\ -i & i \end{bmatrix}\begin{bmatrix} \omega_1 & \omega_2 \\ \omega_1^\rho & \omega_2^\rho \end{bmatrix}\begin{bmatrix} a \\ b \end{bmatrix}.$$

Thus $\mathfrak{c}$ is the image of $\mathbf{Z}^2$ under $2^{-1}\begin{bmatrix} 1 & 1 \\ -i & i \end{bmatrix}\begin{bmatrix} \omega_1 & \omega_2 \\ \omega_1^\rho & \omega_2^\rho \end{bmatrix}$, and so $\mathrm{vol}(\mathbf{C}/\mathfrak{c}) = |\mathrm{Im}(\omega_1\omega_2^\rho)|$, whose square we denote by $d(\mathfrak{c})$. Thus $d(\mathfrak{c})^{1/2}$ is a positive rational number times $|d_K|^{1/2}$, where d_K is the discriminant of K. In particular, if $\mathfrak{c}$ is a fractional ideal in K, then $d(\mathfrak{c}) = N(\mathfrak{c})^2|d_K|/4$.

To study the convergence of (7.1), for $0 < n \in \mathbf{Z}$ consider the parallelogram P_n on the complex plane whose vertices are $\pm(n\omega_1 + n\omega_2)$ and $\pm(n\omega_1 - n\omega_2)$. We easily see that there are exactly $8n$ points $\xi = a\omega_1 + b\omega_2$ of $\mathfrak{c}$ lying on the sides of P_n. Take $\sigma \in \mathbf{R},\, > 0$, so that the circle $|z| = \sigma$ is inside P_1. Then $|\xi|^2 \geq n^2\sigma^2$ for such a ξ lying on the sides of P_n, and so $\sum_{0\neq\xi\in\mathfrak{c}} |\xi|^{-2s} \leq \sigma^{-2s}\sum_{n=1}^{\infty} 8n \cdot n^{-2s}$, which is convergent for $\mathrm{Re}(s) > 1$. Therefore, given α and $\mathfrak{b}$, we take $\mathfrak{c}$ containing $\alpha + \mathfrak{b}$, and find that (7.1) defines a holomorphic function on the half plane $\mathrm{Re}(s) > 1$.

The bilinear form $(x, y) \mapsto {}^t xy$ on $\mathbf{R}^2 \times \mathbf{R}^2$ can be written $(w', w) \mapsto \mathrm{Re}(w'\overline{w})$ for $(w', w) \in \mathbf{C} \times \mathbf{C}$. The restriction of this to $K \times K$ is the bilinear form $(\xi', \xi) \mapsto 2^{-1}\mathrm{Tr}_{K/\mathbf{Q}}(\xi'\xi^\rho)$. Therefore if we define $\tilde{L}$ by (2.5), then

$$\tilde{\mathfrak{b}} = \{\xi \in K \mid \mathrm{Tr}_{K/\mathbf{Q}}(\xi^\rho \mathfrak{b}) \subset 2\mathbf{Z}\}. \tag{7.2}$$

(To avoid the factor 2^{-1} or 2, we have to modify the identification of $\mathbf{C}$ with $\mathbf{R}^2$, which will change the measure on $\mathbf{C}$.)

7.2. We now state various two-dimensional analogues of the formulas of Sections 2 and 3. First, the analogue of (2.22) is

$$\Theta_{\mathfrak{b}}^r(z, \alpha) = \sum_{\xi \in \alpha + \mathfrak{b}} \xi^r \mathbf{e}(|\xi|^2 z/2) \qquad (0 \le r \in \mathbf{Z},\ z \in H). \tag{7.3}$$

The convergence of this series can be seen as follows. Take $\mathfrak{c}$, P_n, and σ as above; take also $T \in \mathbf{R}$, > 0, so that P_1 is inside the circle $|z| = T$. Then

$$\sum_{0 \ne \xi \in \alpha + \mathfrak{b}} \big|\xi^r \mathbf{e}(|\xi|^2 z/2)\big| \le \sum_{n=1}^{\infty} 8n(nT)^r \exp(-\pi n^2 \sigma^2 y), \tag{7.3a}$$

which is clearly convergent. Thus (7.3) defines a holomorphic function on H.

Next, taking "the square" of formula (2.24), we find that

$$\int_{\mathbf{C}} \mathbf{e}(|w_1|^2 z/2)\mathbf{e}\big((-w_1\overline{w} - \overline{w}_1 w)/2\big)\mathbf{d}w_1 = iz^{-1}\mathbf{e}(-|w|^2 z^{-1}/2), \tag{7.4}$$

where w_1 and w are variables on $\mathbf{C}$, $\mathbf{d}w_1 = du_1 dv_1$ for $w_1 = u_1 + iv_1$, and $z \in H$. Applying $(\partial/\partial\overline{w})^r$ to this equality, we obtain

$$\int_{\mathbf{C}} w_1^r \mathbf{e}(|w_1|^2 z/2)\mathbf{e}\big((-w_1\overline{w} - \overline{w}_1 w)/2\big)\mathbf{d}w_1 = iz^{-r-1}w^r \mathbf{e}(-|w|^2 z^{-1}/2). \tag{7.5}$$

(See Theorem A1.4 of the Appendix.) Notice that this is so for every $r > 0$, whereas the action of $(\partial/\partial t)^r$ on (2.24) for $r > 1$ does not give a formula as simple as (2.25).

Take $\mathbf{C}/\mathfrak{b}$ to be $\mathbf{R}^2/L$ in (2.9) and take $f(w) = w^r \mathbf{e}(-|w|^2 z^{-1}/2)$. Then

$$\sum_{b \in \mathfrak{b}} f(\alpha + b) = \sum_{b \in \mathfrak{b}} (\alpha + b)^r \mathbf{e}\big(-|\alpha + b|^2 z^{-1}/2\big) = \Theta_{\mathfrak{b}}^r(-z^{-1}, \alpha).$$

Substituting $-z^{-1}$ for z in (7.5), we find that $\hat{f}(w) = i(-z)^{r+1}w^r \mathbf{e}(|w|^2 z/2)$, and so the right-hand side of (2.9) is

$$i(-z)^{r+1} \sum_{\xi \in \tilde{\mathfrak{b}}} \xi^r \mathbf{e}(|\xi|^2 z/2)\mathbf{e}\big(2^{-1}\mathrm{Tr}_{K/\mathbf{Q}}(\xi\alpha^\rho)\big).$$

Take a $\mathbf{Z}$-lattice $\mathfrak{a}$ in K so that $\mathfrak{a} \subset \tilde{\mathfrak{b}}$ and $\mathrm{Tr}_{K/\mathbf{Q}}(\mathfrak{a}\alpha^\rho) \subset 2\mathbf{Z}$. Then (2.9) gives

$$d(\mathfrak{b})^{1/2}\Theta_{\mathfrak{b}}^r(-z^{-1}, \alpha) = i(-z)^{r+1} \sum_{\beta \in \tilde{\mathfrak{b}}/\mathfrak{a}} \mathbf{e}\big(2^{-1}\mathrm{Tr}_{K/\mathbf{Q}}(\beta\alpha^\rho)\big)\Theta_{\mathfrak{a}}^r(z, \beta). \tag{7.6}$$

We now apply Theorem 3.2 to the present setting by taking $f(z) = \Theta^r_{\mathfrak{b}^\rho}(z, \alpha^\rho)$. We have $D(s, f) = 2^s L^r(s - r/2; \alpha, \mathfrak{b})$. Define $f^\#$ by (3.2) with $N = 1$ and $k = r + 1$. We obtain $f^\#$ from (7.6) and find, with a sufficiently small $\mathfrak{a}$, that

$$(7.7)\qquad D(s, f^\#) = d(\mathfrak{b})^{-1/2} i^{-r} 2^s \sum_{\beta \in \tilde{\mathfrak{b}}^\rho/\mathfrak{a}} \mathbf{e}\big(2^{-1}\mathrm{Tr}_{K/\mathbf{Q}}(\beta\alpha)\big) L^r(s - r/2; \beta, \mathfrak{a}).$$

Theorem 7.3. *The product $\pi^{-s}\Gamma(s + r/2)L^r(s; \alpha, \mathfrak{b})$ can be continued to the whole s-plane as a meromorphic function, which is entire if $r > 0$. If $r = 0$, the product $\pi^{-s}\Gamma(s)L^0(s; \alpha, \mathfrak{b})$ is an entire function plus $d(\mathfrak{b})^{-1/2}/(s - 1) - \delta(\alpha/\mathfrak{b})/s$, where $\delta(\alpha/\mathfrak{b}) = 1$ if $\alpha \in \mathfrak{b}$ and $\delta(\alpha/\mathfrak{b}) = 0$ otherwise.*

This is an immediate consequence of Theorem 3.2.

Substituting $s + r/2$ for s in (3.4b) and employing (7.7), we obtain

$$(7.8)\qquad L^r(1 - s; \alpha, \mathfrak{b}) = \pi^{1-2s}\frac{\Gamma(s + r/2)}{\Gamma(1 - s + r/2)} d(\mathfrak{b})^{-1/2} i^{-r} \sum_{\beta \in \tilde{\mathfrak{b}}^\rho/\mathfrak{a}} \mathbf{e}\big(2^{-1}\mathrm{Tr}_{K/\mathbf{Q}}(\beta\alpha)\big) L^r(s; \beta, \mathfrak{a})$$

with a sufficiently small $\mathfrak{a}$.

8. Nearly holomorphic modular forms

8.1. In later sections, we will investigate the values of $L^r(s; \alpha, \mathfrak{b})$ at some integer values of s, which are, in most cases, "singular values" of certain nonholomorphic modular forms called *nearly holomorphic*. Thus we have to deal with both holomorphic and nonholomorphic functions. We start with a discussion of differential operators on H involving $\mathrm{Im}(z)$. First, for a complex variable $z = x + iy$ we put, as usual,

$$\frac{\partial}{\partial z} = \frac{1}{2}\left(\frac{\partial}{\partial x} - i\frac{\partial}{\partial y}\right), \qquad \frac{\partial}{\partial \overline{z}} = \frac{1}{2}\left(\frac{\partial}{\partial x} + i\frac{\partial}{\partial y}\right).$$

We note that

$$(8.1)\qquad (\partial/\partial z)(f \circ g) = (\partial f/\partial z)(g(z)) \cdot dg/dz$$

if g is holomorphic (even if f is nonholomorphic). In particular, we have, for a differentiable function f,

$$(8.2)\qquad (\partial/\partial z)f(\alpha z) = j_\alpha(z)^{-2}(\partial f/\partial z)(\alpha z) \quad \text{for} \quad \alpha \in GL_2(\mathbf{C}).$$

To make our formulas short and transparent, we put

$$(8.3)\qquad \eta(z) = z - \overline{z} = 2iy \qquad (z \in H).$$

Then clearly $(\partial/\partial z)\eta^m = m\eta^{m-1}$ for every $m \in \mathbf{Z}$ and $\eta(\alpha z) = |j_\alpha(z)|^{-2}\eta(z)$ for every $\alpha \in G_{\mathfrak{a}+}$.

For $k \in \mathbf{Z}$ and $0 \le p \in \mathbf{Z}$ we define differential operators δ_k, δ_k^p, D_k, and D_k^p acting on differentiable functions f defined on the whole of H or its open subset by

$$\delta_k f = \eta^{-k}\frac{\partial}{\partial z}(\eta^k f) = \left(\frac{k}{\eta} + \frac{\partial}{\partial z}\right) f, \tag{8.4}$$

$$\delta_k^{p+1} = \delta_{k+2p}\delta_k^p, \quad \delta_k^1 = \delta_k, \quad \delta_k^0 = 1, \tag{8.5}$$

$$D_k = (2\pi i)^{-1}\delta_k, \qquad D_k^p = (2\pi i)^{-p}\delta_k^p. \tag{8.6}$$

In particular, $D_0 f = (2\pi i)^{-1}\partial f/\partial z$. As the most basic property of these operators we have

$$\delta_k^p(f\|_k\,\alpha) = (\delta_k^p f)\|_{k+2p}\,\alpha \quad \text{for every} \quad \alpha \in G_{\mathbf{a}+}, \tag{8.7}$$

and consequently $D_k^p(f\|_k\,\alpha) = (D_k^p f)\|_{k+2p}\,\alpha$ for every $\alpha \in G_{\mathbf{a}+}$. If $p = 1$, (8.7) becomes

$$\delta_k(f\|_k\,\alpha) = (\delta_k f)\|_{k+2}\,\alpha \quad \text{for every} \quad \alpha \in G_{\mathbf{a}+}, \tag{8.8}$$

and (8.7) follows from (8.8) by induction on p. The proof of (8.8) goes as follows. We have

$$\begin{aligned} \delta_k(f\|_k\,\alpha) &= \eta^{-k}(\partial/\partial z)\big(\eta^k j_\alpha^{-k} f(\alpha z)\big) = \eta^{-k}(\partial/\partial z)\big(\eta(\alpha z)^k \overline{j_\alpha}^{-k} f(\alpha z)\big) \\ &= \eta^{-k}\overline{j_\alpha}^{-k}(\partial/\partial z)\big((\eta^k f)\circ\alpha\big) = \eta^{-k}\overline{j_\alpha}^{-k} j_\alpha^{-2}\{(\partial/\partial z)(\eta^k f)\}(\alpha z) \\ &= j_\alpha^{-k-2}\eta(\alpha z)^{-k}\{(\partial/\partial z)(\eta^k f)\}(\alpha z) = (\delta_k f)\|_{k+2}\alpha, \end{aligned}$$

which gives (8.8). We have also

$$\delta_{k+\ell}(fg) = g\delta_k f + f\delta_\ell g, \quad \delta_{k+\ell}^m(fg) = \sum_{\nu=0}^m \binom{m}{\nu}\delta_k^\nu f\cdot\delta_\ell^{m-\nu} g, \tag{8.9}$$

$$f^{-m}\delta_{mk}(f^m) = mf^{-1}\delta_k f, \tag{8.10}$$

$$\begin{aligned} \delta_k^p f &= \eta^{1-k-p}(\partial/\partial z)^p(\eta^{k+p-1}f) \\ &= \sum_{\nu=0}^p \binom{p}{\nu}\frac{\Gamma(k+p)}{\Gamma(k+\nu)}\eta^{\nu-p}f^{(\nu)} \qquad (0 \le p \in \mathbf{Z}). \end{aligned} \tag{8.11}$$

These except (8.11) follow immediately from our definition of δ_k and δ_k^p. To prove (8.11), put $f^{(\nu)} = (\partial/\partial z)^\nu f$ and denote by $\varepsilon_k^p f$ the last sum over ν. Then

$$\begin{aligned} &\delta_{k+2p}\varepsilon_k^p f = \eta^{-k-2p}(\partial/\partial z)(\eta^{k+2p}\varepsilon_k^p f) \\ &= \eta^{-k-2p}\sum_{\nu=0}^p \binom{p}{\nu}\frac{\Gamma(k+p)}{\Gamma(k+\nu)}(\partial/\partial z)(\eta^{k+p+\nu}f^{(\nu)}) \\ &= \sum_{\nu=0}^p \binom{p}{\nu}\frac{\Gamma(k+p)}{\Gamma(k+\nu)}(k+p+\nu)\eta^{\nu-p-1}f^{(\nu)} + \sum_{\nu=0}^p \binom{p}{\nu}\frac{\Gamma(k+p)}{\Gamma(k+\nu)}\eta^{\nu-p}f^{(\nu+1)}. \end{aligned}$$

We can easily verify that this equals $\varepsilon_k^{p+1}f$. Thus $\varepsilon_k^{p+1} = \delta_{k+2p}\varepsilon_k^p$. Clearly $\delta_k = \varepsilon_k^1$, and so we obtain $\varepsilon_k^p = \delta_k^p$ by induction on p. Now by the Leibniz rule we easily see that $\eta^{1-k-p}(\partial/\partial z)^p(\eta^{k+p-1}f) = \varepsilon_k^p f$, which completes the proof of (8.11).

From (8.9) and (8.10) we obtain

$$\delta_{ak+b\ell}(f^a g^b) = a f^{a-1} g^b \delta_k f + b f^a g^{b-1} \delta_\ell g \qquad (a,\, b \in \mathbf{Z}). \tag{8.12}$$

The weight k is an integer in the whole book except in Section 16, in which a more general case will be considered.

8.2. We now introduce the notion of nearly holomorphic forms. For a congruence subgroup Γ of $\Gamma(1)$, $0 \le t \in \mathbf{Z}$, and $0 \le k \in \mathbf{Z}$ we denote by $\mathscr{N}_k^t(\Gamma)$ the set of all $\mathbf{C}$-valued functions f on H satisfying the following two conditions:

(8.13a) $f\|_k\gamma = f$ for every $\gamma \in \Gamma$;

(8.13b) For every $\alpha \in G_+$ we have $(f\|_k\alpha)(z) = \sum_{\nu=0}^{t} \eta^{-\nu} \sum_{n=0}^{\infty} c_{\nu\alpha n}\mathbf{e}(nz/N_\alpha)$ with $c_{\nu\alpha n} \in \mathbf{C}$ and $0 < N_\alpha \in \mathbf{Z}$.

Since $G_+ = \Gamma(1)P_+$ as noted in (1.15), the group G_+ in (8.13a) can be replaced by $\Gamma(1)$. Therefore, if $\Gamma = \Gamma(1)$, for example, we need (8.13b) only for $\alpha = 1$. Also, for a subfield Φ of $\mathbf{C}$ we denote by $\mathscr{N}_k^t(\Phi,\, \Gamma)$ the set of all $f \in \mathscr{N}_k^t(\Gamma)$ such that $f(z) = \sum_{a=0}^{t}(2\pi y)^{-a}\sum_{n=0}^{\infty} c_{an}\mathbf{e}(nz/N)$ with $c_{an} \in \Phi$ and $0 < N \in \mathbf{Z}$. We then put

$$\mathscr{N}_k(\Gamma) = \bigcup_{t=0}^{\infty} \mathscr{N}_k^t(\Gamma), \qquad \mathscr{N}_k(\Phi,\, \Gamma) = \bigcup_{t=0}^{\infty} \mathscr{N}_k^t(\Phi,\, \Gamma), \tag{8.14}$$

$$\mathscr{N}_k^t = \textstyle\bigcup_\Gamma \mathscr{N}_k^t(\Gamma), \qquad \mathscr{N}_k^t(\Phi) = \bigcup_\Gamma \mathscr{N}_k^t(\Phi,\, \Gamma), \tag{8.15}$$

$$\mathscr{N}_k = \textstyle\bigcup_\Gamma \mathscr{N}_k(\Gamma), \qquad \mathscr{N}_k(\Phi) = \bigcup_\Gamma \mathscr{N}_k(\Phi,\, \Gamma), \tag{8.16}$$

where Γ runs over all congruence subgroups of $SL_2(\mathbf{Z})$. We call the elements of $\mathscr{N}_k(\Gamma)$ **nearly holomorphic modular forms of weight** k with respect to Γ. Clearly the symbol $\mathscr{N}_k^0$ is the same as $\mathscr{M}_k$. As to other basic facts on the forms of this type such as their characterization in terms of Lie algebras, the reader is referred to [S87], [S00], and also earlier relevant papers of the author mentioned there.

There is a well known element E_2 of $\mathscr{N}_2^1(\mathbf{Q},\, \Gamma(1))$, which is given by

$$E_2(z) = \frac{1}{8\pi y} - \frac{1}{24} + \sum_{n=1}^{\infty}\Bigl(\sum_{0<d|n} d\Bigr)\mathbf{e}(nz). \tag{8.17}$$

Indeed, taking the logarithmic derivative of (1.12), we easily find that

$$\Delta^{-1} D_{12}\Delta = -24E_2. \tag{8.18}$$

Therefore by (8.8), $E_2\|_2\gamma = E_2$ for every $\gamma \in \Gamma(1)$, and so from (8.17) we see that E_2 satisfies (8.13a, b) with $\Gamma = \Gamma(1)$. In the next section we will show that E_2 is an Eisenstein series.

Lemma 8.3. *For every subfield Φ of $\mathbf{C}$ and every congruence subgroup Γ of $\Gamma(1)$ the following assertions hold:*

(i) *The operator* D_k^p *sends* $\mathscr{N}_k^t(\Phi, \Gamma)$ *into* $\mathscr{N}_{k+2p}^{t+p}(\Phi, \Gamma)$; *in particular, it sends* $\mathscr{M}_k(\Phi, \Gamma)$ *into* $\mathscr{N}_{k+2p}^p(\Phi, \Gamma)$.

(ii) *Every element* $f(z)$ *of* $\mathscr{N}_k(\Phi, \Gamma)$ *can be written uniquely in the form* $f(z) = \sum_{0 \le a \le k/2} E_2^a h_a$ *with* $h_a \in \mathscr{M}_{k-2a}(\Phi, \Gamma)$. *Consequently,* $\mathscr{N}_1 = \{0\}$.

(iii) $D_k f + 2kE_2 f \in \mathscr{M}_{k+2}(\Phi, \Gamma)$ *for every* $f \in \mathscr{M}_k(\Phi, \Gamma)$ *and* $D_k f + 2kE_2 f \in \mathscr{S}_{k+2}(\Phi, \Gamma)$ *for every* $f \in \mathscr{S}_k(\Phi, \Gamma)$.

(iv) $\mathscr{N}_k^t(\Phi)$ *is stable under the map* $f \mapsto f\|_k\alpha$ *for every* $\alpha \in G_+$, *provided* $\mathbf{Q}_{ab} \subset \Phi$.

PROOF. From the definition of D_k and (8.8), we easily see that (i) is true if $p = 1$. Then the general case can be proved by induction on p. To prove (ii), let $f \in \mathscr{N}_k(\Phi, \Gamma)$. Then $f(z) = \sum_{a=0}^t (2\pi y)^{-a} f_a(z)$ with holomorphic functions f_a on H. For $\gamma \in G_+$ we have $(y \circ \gamma)^{-a} = |j_\gamma|^{2a} y^{-a}$ and $\overline{j}_\gamma = j_\gamma - 2i\varepsilon c_\gamma y$ with $\varepsilon = \det(\gamma)^{-1/2}$, and so

$$f\|_k\gamma = \sum_{a=0}^{t} (2\pi y)^{-a} j_\gamma^{a-k} (j_\gamma - 2i\varepsilon c_\gamma y)^a (f_a \circ \gamma), \tag{8.19}$$

which can be written $\sum_{a=0}^t (2\pi y)^{-a} g_{a\gamma}$ with holomorphic functions $g_{a\gamma}$. Viewing this as a polynomial in y^{-1} and comparing the coefficients of y^{-t}, we obtain $g_{t\gamma} = j_\gamma^{2t-k}(f_t \circ \gamma)$, that is, $g_{t\gamma} = f_t\|_{k-2t}\gamma$, because the function y^{-1} is an algebraically independent variable over the field of meromorphic functions on H. In particular, $g_{t\gamma} = f_t$ if $\gamma \in \Gamma$, and so $f_t\|_{k-2t}\gamma = f_t$ for $\gamma \in \Gamma$. By condition (8.13b), $g_{t\gamma}$ has a Fourier expansion finite at ∞, this shows that $f_t \in \mathscr{M}_{k-2t}(\Gamma)$. Therefore, if $t > 0$ and $f_t \neq 0$, then $k \ge 2t$. From the definition of $\mathscr{N}_k(\Phi, \Gamma)$ we see that $f_t \in \mathscr{M}_{k-2t}(\Phi, \Gamma)$. Then $f - 4^t E_2^t f_t$ is an element of $\mathscr{N}_k(\Phi, \Gamma)$ of the form $\sum_{a=0}^{t-1} (2\pi y)^a \ell_a(z)$ with holomorphic ℓ_a. Subtracting $4^{t-1} E_2^{t-1} \ell_{t-1}$ from this sum (if $t > 1$), and repeating the same procedure, we eventually find that $f = \sum_{a=0}^t E_2^a h_a$ with $h_a \in \mathscr{M}_{k-2a}(\Phi, \Gamma)$. Viewing this as a polynomial in y^{-1}, we easily find that the expression is unique. To prove (iii), put $g = D_k f + 2kE_2 f$. Then we see that g is holomorphic, and belongs to $\mathscr{N}_{k+2}^1(\Phi, \Gamma)$ by (i), and so $g \in \mathscr{M}_{k+2}(\Phi, \Gamma)$. If $f \in \mathscr{S}_k$, then for every $\alpha \in \Gamma(1)$, we have $g\|_{k+2}\alpha = D_k(f\|_k\alpha) + 2kE_2 \cdot (f\|_k\alpha)$, whose constant term is 0. Thus $g \in \mathscr{S}_{k+2}$. As for (iv), from (8.13a,b) and (8.19) we see that $\mathscr{N}_k^t$ is stable under the map $f \mapsto f\|_k\alpha$ for every $\alpha \in G_+$. Now (iv) is clear for $\alpha \in P_+$, and so in view of (1.15), it is sufficient to prove the case $\alpha \in \Gamma(1)$. Given $f \in \mathscr{N}_k(\Phi, \Gamma)$, by (ii) we can put $f = \sum_{0 \le a \le k/2} E_2^a h_a$ with $h_a \in \mathscr{M}_{k-2a}(\Phi, \Gamma)$. If $\alpha \in \Gamma(1)$, then $f\|_k\alpha = \sum_a E_2^a \cdot (h_a\|_k\alpha)$. Suppose $\mathbf{Q}_{ab} \subset \Phi$; then by Theorem 1.5, $h_a\|_k\alpha \in \mathscr{M}_{k-2a}(\Phi)$, and so $f\|_k\alpha \in \mathscr{N}_k^t(\Phi)$ as expected. This completes the proof.

The principle stated in (iii) above often gives explicit formulas for $D_k f$ for explicitly given f. We will state some such formulas in the following two sections, and add a few comments in §16.8.

CHAPTER IV

EISENSTEIN SERIES

9. Fourier expansions of Eisenstein series

9.1. For $0 \le k \in \mathbf{Z}$, a positive integer N, and $(p, q) \in \mathbf{Z}^2$ we put

$$(9.1)\quad \mathfrak{E}_k^N(z, s; p, q) = \mathrm{Im}(z)^s \sum_{(m,n)} (mz+n)^{-k}|mz+n|^{-2s} \qquad (z \in H,\ s \in \mathbf{C}),$$

$$(9.2)\quad \mathfrak{E}_k^N(z; p, q) = \mathfrak{E}_k^N(z, 0; p, q) \qquad (k > 0),$$

where (m, n) runs over $\mathbf{Z}^2$ under the condition $0 \ne (m, n) \equiv (p, q) \pmod{N\mathbf{Z}^2}$. We showed in §7.1 that the sum $\sum |a\omega_1 + b\omega_2|^{-2s}$ with $0 \ne (a, b) \in \mathbf{Z}^2$ is convergent for $\mathrm{Re}(s) > 1$ if $\mathbf{Z}\omega_1 + \mathbf{Z}\omega_2$ is a lattice in $\mathbf{C}$. Applying this fact to the lattice $\mathbf{Z}z + \mathbf{Z}$ with $z \in H$, we see that the series of (9.1) is convergent for $\mathrm{Re}(2s)+k > 2$. Moreover we will prove in Theorem 9.7 that it can be continued to a meromorphic function in s on the whole plane satisfying a functional equation. If $k > 0$ in particular, it can be continued to an entire function of s. Thus the function of (9.2) is meaningful.

For $\gamma \in \Gamma(1)$ and $(m', n') = (m, n)\gamma$ we have $m'z + n' = (m\gamma(z) + n)j_\gamma(z)$, and so we can easily verify that

$$(9.3)\quad \mathfrak{E}_k^N(z, s; (p, q)\gamma) = \mathfrak{E}_k^N(\gamma z, s; p, q)j_\gamma(z)^{-k} \quad \text{for every} \quad \gamma \in \Gamma(1).$$

In particular, we have

$$(9.3a)\quad \mathfrak{E}_k^N(z, s; p, q)\|_k\gamma = \mathfrak{E}_k^N(z, s; p, q) \quad \text{for every} \quad \gamma \in \Gamma(N),$$

$$(9.3b)\quad \mathfrak{E}_k^N(z, s; -p, -q) = (-1)^k\mathfrak{E}_k^N(z, s; p, q),$$

$$(9.3c)\quad \mathfrak{E}_k^N(z, s; q, -p) = z^{-k}\mathfrak{E}_k^N(-z^{-1}, s; p, q),$$

$$(9.3d)\quad h^{-k-2s}\mathfrak{E}_k^N(z, s; p, q) = \mathfrak{E}_k^{hN}(z, s; hp, hq) \qquad (0 < h \in \mathbf{Z}).$$

Now the Fourier expansion of $\mathfrak{E}_k^N(z; p, q)$ can be given as follows:

$$(9.4)\quad (-2\pi i)^{-k}N^k\Gamma(k)\mathfrak{E}_k^N(z; p, q)$$
$$= A + B + \sum_{n=1}^{\infty} \sum_{0<m\in p+N\mathbf{Z}} n^{k-1}\mathbf{e}\big(n(mz+q)/N\big)$$
$$+ \sum_{n=1}^{\infty} \sum_{0<m\in -p+N\mathbf{Z}} (-1)^k n^{k-1}\mathbf{e}\big(n(mz-q)/N\big),$$

where A and B are determined by

(9.5a)
$$A = \begin{cases} (-2\pi i)^{-k} N^k \Gamma(k) M_{q,N}^k(0) & \text{if } p \in N\mathbf{Z}, \\ 0 & \text{otherwise}, \end{cases}$$

(9.5b)
$$B = \begin{cases} 2^{-1} M_{p,N}^1(-1/2) & \text{if } k = 1 \text{ and } p \notin N\mathbf{Z}, \\ (4\pi y)^{-1} & \text{if } k = 2, \\ 0 & \text{otherwise}, \end{cases}$$

(9.5c)
$$M_{q,N}^k(s) = \sum_{0 \neq m \in q + N\mathbf{Z}} m^{-k}|m|^{-2s}.$$

The function $M_{q,N}^k$ is the same as what we defined in (3.5).

Such Fourier expansions were obtained in substance by Hecke in [He27], though he did not give constant terms in clear-cut finite forms. The last comment applies even to later authors. The explicit forms of Eisenstein series of "Haupttypus und Nebentypus" belonging to $\Gamma_1(N)$ were given in [He40], under the condition that N is a prime and $k \geq 2$. In such cases, he gave the constant terms by proving (4.26), as we already noted in Section 4. Here we followed the formulation of [S00, Section 18], and we give the details of the proof in §A3.7 of the Appendix.

We observe that $\mathfrak{E}_k^N(z;\, p,\, q)$, if $k \neq 2$, is holomorphic in z, and (9.3) together with (9.4) shows that (1.9b) is satisfied, and therefore it belongs to $\mathscr{M}_k(\Gamma(N))$. For $k = 2$, we have the term $(4\pi y)^{-1}$ as noted in (9.5b), and so (8.13b) is satisfied with $t = 1$. Thus $\mathfrak{E}_2^N(z;\, p,\, q)$ is nearly holomorphic and belongs to $\mathscr{N}_2^1(\Gamma(N))$.

Put $\Phi_N = \mathbf{Q}(\mathbf{e}(1/N))$. Then the function $(2\pi i)^{-k}\mathfrak{E}_k^N(z;\, p,\, q)$ belongs to $\mathscr{M}_k(\Phi_N)$ or $\mathscr{N}_2^1(\Phi_N)$ according as $k \neq 2$ or $k = 2$. Indeed, (9.4) shows clearly that the Fourier coefficients, except the constant term, belong to Φ_N. As for the constant term, we first note that

(9.6a) $\quad M_{q,N}^k(s) = 0 \quad \text{if} \quad q \in N\mathbf{Z} \quad \text{and} \quad k \notin 2\mathbf{Z},$

(9.6b) $\quad M_{q,N}^k(s) = 2N^{-k-2s}\zeta(2s+k) \quad \text{if} \quad q \in N\mathbf{Z} \quad \text{and} \quad k \in 2\mathbf{Z},$

Thus for $q \in N\mathbf{Z}$, the value $M_{q,N}^k(0)$ is 0 if $k \notin 2\mathbf{Z}$ and $2N^{-k}\zeta(k)$ if $k \in 2\mathbf{Z}$, which belongs to $\pi^k\mathbf{Q}^\times$ by (4.25). For $q \notin N\mathbf{Z}$ the value $M_{q,N}^k(0)$ can be obtained by (3.9). As for $M_{p,N}^k(-1/2)$ that appears in (9.5b), it is given by (3.8). From these formulas we see that the constant term belongs to Φ_N, which proves the desired fact on $(2\pi i)^{-k}\mathfrak{E}_k^N(z;\, p,\, q)$.

9.2. If $N = 1$, the Fourier expansion of our Eisenstein series has a well known clear-cut form. For $2 \leq k \in 2\mathbf{Z}$ we put

(9.7)
$$E_k(z) = 2^{-1}(2\pi i)^{-k}\Gamma(k)\mathfrak{E}_k^1(z;\, 0,\, 0).$$

For $k = 2$ the function E_2 is exactly the nonholomorphic form of (8.17). For $k > 2$ we have

$$E_k(z) = \frac{(k-1)!}{(2\pi i)^k} \cdot \zeta(k) + \sum_{n=1}^{\infty} \mathbf{e}(nz)\Big\{ \sum_{0<d|n} d^{k-1} \Big\}. \tag{9.8}$$

Thus E_k is normalized in the sense that the coefficient of $\mathbf{e}(z)$ is 1. Let A_k denote the constant term of $E_k(z)$. By (4.25) we have

$$A_k = -B_k/(2k). \tag{9.8a}$$

From (4.12) we see that $(-1)^m A_{2m} > 0$. We note here the first four values of A_k:

$$A_4 = 1/240, \quad A_6 = -1/504, \quad A_8 = 1/480, \quad A_{10} = -1/264. \tag{9.8b}$$

We add here several more examples of Lemma 8.3 (ii), (iii), which will be needed later:

$$D_2E_2 = (5/6)E_4 - 2E_2^2, \tag{9.9a}$$

$$D_4E_4 = (7/10)E_6 - 8E_2E_4, \tag{9.9b}$$

$$D_6E_6 = (10/21)E_8 - 12E_2E_6, \tag{9.9c}$$

$$D_2^2E_2 = (7/12)E_6 - 10E_2E_4 + 8E_2^3. \tag{9.9d}$$

The first three formulas can be verified by checking the nonholomorphic and constant terms, as $\dim \mathscr{M}_k\big(\Gamma(1)\big) = 1$ for even k such that $4 \le k \le 10$. Then the last one follows from these by applying D_2 to (9.9a). In the next section we will give a recurrence formula by which we can express E_k for $k \ge 8$ as a polynomial in E_h with $h < k$. Therefore we obtain inductively the formula for $D_k^n E_k$ as a $\mathbf{Q}$-linear combination of the monomials $E_2^a E_4^b E_6^c$ such that $2a + 4b + 6c = k + 2n$. The formula for $D_k^n \mathfrak{E}_k^N(z, s;\, p, q)$ will be given in (9.12).

9.3. We will discuss explicit examples of Eisenstein series belonging to $\Gamma_1(N)$ in the next section. Here we mention two types, which depend on a primitive Dirichlet character χ modulo N. Assuming that $N > 1$ and $\chi(-1) = (-1)^k$, put

$$\varphi_{k,\chi}^{\ell}(z) = \frac{N^k \Gamma(k)}{2(-2\pi i)^k} \sum_{p=1}^{N-1} \chi(p) \mathfrak{E}_k^N(z;\, p, 0), \tag{9.10a}$$

$$\varphi_{k,\chi}^{r}(z) = \frac{N^k \Gamma(k)}{2G(\chi)(-2\pi i)^k} \sum_{q=1}^{N-1} \chi(q) \mathfrak{E}_k^N(z;\, 0, q), \tag{9.10b}$$

where $G(\chi)$ is the Gauss sum of (4.20). These have the following expansions:

$$\varphi_{k,\chi}^{\ell}(z) = C + \sum_{n=1}^{\infty} \mathbf{e}(nz/N) \sum_{0<d|n} \chi(n/d) d^{k-1}, \tag{9.11a}$$

$$C = \begin{cases} \dfrac{-1}{2N}\displaystyle\sum_{t=1}^{N-1}\chi(t)t & (k=1), \\ 0 & (k>1), \end{cases}$$

(9.11b) $$\varphi^r_{k,\chi}(z) = \frac{N^k\Gamma(k)}{G(\chi)(-2\pi i)^k}L(k,\,\chi) + \sum_{n=1}^{\infty}\mathbf{e}(nz)\sum_{0<d|n}\overline{\chi}(d)d^{k-1}.$$

The nonconstant terms of these can be verified in a straightforward way in view of (4.21). To obtain the constant term of $\varphi^r_{k,\chi}$, we note that $\sum_{q=1}^{N}\chi(q)M^k_{q,N}(s) = 2L(2s+k,\,\chi)$, and so $\sum_{q=1}^{N}\chi(q)M^k_{q,N}(0) = 2L(k,\,\chi)$. We can apply (4.16a), (4.26), (4.32–36) to this constant term. As for the constant term of $\varphi^\ell_{k,\chi}$, it is clearly 0 if $k>1$. For $k=1$ we have $2^{-1}\sum_p\chi(p)M^1_{p,N}(-1/2) = L(0,\,\chi)$, which is given by (4.27).

From (9.3) we obtain the first three of the following formulas:

(9.11c) $$\varphi^\ell_{k,\chi}(-z^{-1})(-z)^{-k} = G(\chi)\varphi^r_{k,\chi}(z),$$

(9.11d) $$\varphi^\ell_{k,\chi}\|_k\gamma = \chi(d_\gamma)\varphi^\ell_{k,\chi} \quad \text{for every} \quad \gamma\in\Gamma^0(N),$$

(9.11e) $$\varphi^r_{k,\chi}\|_k\gamma = \chi(a_\gamma)\varphi^r_{k,\chi} \quad \text{for every} \quad \gamma\in\Gamma_0(N),$$

(9.11f) If $\psi(z) = \varphi^\ell_{k,\chi}(Nz)$, then $\psi\|_k\gamma = \chi(d_\gamma)\psi$ for every $\gamma\in\Gamma_0(N)$,

(9.11g) If $\omega(z) = \varphi^r_{k,\chi}(z/N)$, then $\omega\|_k\gamma = \chi(a_\gamma)\omega$ for every $\gamma\in\Gamma^0(N)$,

(9.11h) $$\varphi^\ell_{1,\chi}(Nz) = \varphi^r_{1,\overline{\chi}}(z).$$

The last formula follows from (9.10a, b) and (9.11f, g) follow from (9.11d, e).

9.4. The explicit action of the operator D^n_k of (8.6) on Eisenstein series is given by

(9.12) $$D^n_k\mathfrak{E}^N_k(z,\,s;\,p,\,q) = (-4\pi)^{-n}\frac{\Gamma(k+s+n)}{\Gamma(k+s)}\mathfrak{E}^N_{k+2n}(z,\,s-n;\,p,\,q).$$

To prove this, we first note

(9.13) $$\delta^n_k(y^s) = (2i)^{-n}\frac{\Gamma(k+s+n)}{\Gamma(k+s)}\cdot y^{s-n} \qquad (s\in\mathbf{C}),$$

which can be verified by induction on n, or by (8.11). Since $y^s\|_k\alpha = y^s(cz+d)^{-k}|cz+d|^{-2s}$ for $\alpha = \begin{bmatrix} a & b \\ c & d\end{bmatrix}\in G^1_{\mathbf{a}}$, we have, by (9.13) and (8.8),

$$\begin{aligned}\delta^n_k\bigl(y^s(cz+d)^{-k}|cz+d|^{-2s}\bigr) &= \delta^n_k(y^s\|_k\alpha) = (\delta^n_k y^s)\|_{k+2n}\,\alpha \\ &= (2i)^{-n}\Gamma(k+s)^{-1}\Gamma(k+s+n)y^{s-n}\|_{k+2n}\,\alpha \\ &= (2i)^{-n}\Gamma(k+s)^{-1}\Gamma(k+s+n)y^{s-n}(cz+d)^{-k-2n}|cz+d|^{2n-2s}.\end{aligned}$$

Therefore termwise differentiation proves (9.12) for sufficiently large $\mathrm{Re}(s)$, and even for every s, because both sides are holomorphic in s on the whole $\mathbf{C}$.

9.5. We now consider the explicit Fourier expansion of $\mathfrak{E}_k^N(z, s; p, q)$ at $s = 1-k$. If $k = 1$, the function is $\mathfrak{E}_1^N(z, 0; p, q)$, whose Fourier expansion was given in (9.4). Therefore we consider only the case $k > 1$. The result is as follows:

$$(9.14)\qquad (-2i)^{-k}N\pi^{-1}\mathfrak{E}_k^N(z, 1-k; p, q)$$
$$= A' + B' + \sum_{n=1}^{\infty}\sum_{0<m\in p+N\mathbf{Z}} m^{k-1}\mathbf{e}\big(n(mz+q)/N\big)$$
$$+ \sum_{n=1}^{\infty}\sum_{0<m\in -p+N\mathbf{Z}} (-1)^k m^{k-1}\mathbf{e}\big(n(mz-q)/N\big),$$

$$(9.15\text{a})\qquad A' = \begin{cases} (4\pi y)^{-1}N & \text{if } k = 2 \text{ and } (p, q) \in N\mathbf{Z}^2, \\ 0 & \text{otherwise,} \end{cases}$$

$$(9.15\text{b})\qquad B' = -k^{-1}N^{k-1}B_k(p/N),$$

where $B_k(t)$ is the Bernoulli polynomial of (4.7b). The proof will be given in §A3.8 of the Appendix. Notice that the function of (9.14) for $k = 2$ is nonholomorphic only when $(p, q) \in N\mathbf{Z}^2$, while (9.4) for $k = 2$ is *always* nonholomorphic. We easily see that

$$(9.16)\qquad (-2i)^{-k}\pi^{-1}\mathfrak{E}_k^1(z, 1-k; 0, 0) = 2E_k(z) \qquad (0 < k \in 2\mathbf{Z}).$$

Theorem 9.6. *Denote the function* $\mathfrak{E}_k^N(z, s; p, q)$, *for any fixed* (N, k, p, q) *as in §9.1, simply by* $\mathfrak{E}_k(z, s)$. *Then the following assertions hold:*

(i) $\pi^{-k}\mathfrak{E}_k(z, 0)$ *belongs to* $\mathscr{M}_k(\mathbf{Q}_{\mathrm{ab}})$ *except when* $k = 2$, *in which case it belongs to* $\mathscr{N}_2^1(\mathbf{Q}_{\mathrm{ab}})$.

(ii) $\pi^{-1}\mathfrak{E}_k(z, 1-k)$ *belongs to* $\mathscr{M}_k(\mathbf{Q}_{\mathrm{ab}})$ *except when* $k = 2$, *in which case it belongs to* $\mathscr{N}_2^1(\mathbf{Q}_{\mathrm{ab}})$.

(iii) *Let* n *be an integer such that* $0 \le n \le k-1$. *Then* $\pi^{n-k}\mathfrak{E}_k(z, -n)$ *belongs to* $\mathscr{N}_k^t(\mathbf{Q}_{\mathrm{ab}})$, *where* $t = k-n-1$ *if* $n > k/2$ *or* $n = (k-2)/2$, *and* $t = n$ *in all the remaining cases.*

PROOF. We have already seen (i) in §9.1, which includes (ii) in the case $k = 1$; for $k > 1$, (ii) can be seen from (9.14). To prove (iii), we first assume that $0 \le n \le (k-1)/2$. Putting $\ell = k-2n$, from (9.12) (with ℓ in place of k there) we obtain

$$(9.17\text{a})\qquad \Gamma(\ell+n)\mathfrak{E}_k(z, -n) = (-4\pi)^n\Gamma(\ell)D_\ell^n\mathfrak{E}_\ell(z, 0).$$

Therefore we obtain (iii) in this case by combining (i) with Lemma 8.3 (i). Next suppose $(k-1)/2 < n \le k-1$. Put $m = k-n-1$ and $k-2m = h$. Then $0 \le m < (k-1)/2$, $h > 1$, and (9.12) gives

$$(9.17\text{b})\qquad \Gamma(m+1)\mathfrak{E}_k(z, -n) = (-4\pi)^m D_h^m\mathfrak{E}_h(z, 1-h).$$

Therefore we obtain the desired fact from (ii) and Lemma 8.3 (i).

Theorem 9.7. *For $k \geq 0$, N, and (p, q) as in §9.1 put*

$$Z_k^N(z,\, s;\, p,\, q) = \Gamma(s+k)\pi^{-s}\mathfrak{E}_k^N(z,\, s;\, p,\, q). \tag{9.18}$$

Then this function can be continued as a meromorphic function of s to the whole complex plane. It is an entire function if $k > 0$; if $k = 0$, it is an entire function plus $N^{-2}/(s-1) - \delta/s$, where $\delta = 1$ if $(p, q) \in N\mathbf{Z}^2$ and $\delta = 0$ otherwise. Moreover we have

$$\begin{aligned} &Z_k^N(z,\, 1-k-s;\, p,\, q) \\ &= N^{2s+k-2} \sum_{(a,b)\in\mathbf{Z}^2/N\mathbf{Z}^2} \mathbf{e}\big((qa-pb)/N\big) Z_k^N(z,\, s;\, a,\, b). \end{aligned} \tag{9.19}$$

Remark. In Theorem A3.5 of the Appendix we will investigate the real analyticity of the function in $(z, s) \in H \times \mathbf{C}$. See also §A3.9.

Before proving our theorem, we state a few formulas derived from (9.19). Assuming $k > 0$ and evaluating (9.19) at $s = 0$, we obtain an equality connecting the function of (9.14) to that of (9.4):

$$\begin{aligned} &(-2i)^{-k}N\pi^{-1}\mathfrak{E}_k^N(z,\, 1-k;\, p,\, q) \\ &= N^{-1} \sum_{(a,b)\in\mathbf{Z}^2/N\mathbf{Z}^2} \mathbf{e}\big((qa-pb)/N\big)(-2\pi i)^{-k}N^k\Gamma(k)\mathfrak{E}_k^N(z;\, a,\, b). \end{aligned} \tag{9.20}$$

In the simplest case $N = 1$ we have

$$\begin{aligned} Z_k^1(z,\, s;\, 0,\, 0) &= \Gamma(s+k)\pi^{-s}\mathfrak{E}_k^1(z,\, s;\, 0,\, 0) \\ &= Z_k^1(z,\, 1-k-s;\, 0,\, 0), \end{aligned} \tag{9.21}$$

$$(-2i)^{-k}\pi^{-1}\mathfrak{E}_k^1(z,\, 1-k;\, 0,\, 0) = 2E_k(z). \tag{9.22}$$

PROOF OF THEOREM 9.7. We start with the equality

$$\begin{aligned} &\int_{\mathbf{R}^2} (u\bar{z}+v)^k \exp\big(-\pi t y^{-1}|uz+v|^2\big)\mathbf{e}(-ur-vs)\,du\,dv \\ &\qquad = t^{-k-1}(s\bar{z}-r)^k \exp\big(-\pi t^{-1}y^{-1}|sz-r|^2\big). \end{aligned} \tag{9.23}$$

Here $0 \leq k \in \mathbf{Z}$, $0 < t \in \mathbf{R}$, $z \in H$, $y = \mathrm{Im}(z)$, and u, v, r, s are real variables. To prove this, we first substitute k and $iy^{-1}t$ for r and z in (7.5). After that, put $w_1 = u\bar{z} + v$ and $w = i(s\bar{z} - r)/y$. If the variables u_1, v_1 are determined by $w_1 = u_1 i + v_1$, then $du_1 dv_1 = y \cdot du dv$. Since $w_1\overline{w} + \overline{w}_1 w = 2(ur + vs)$, we obtain (9.23) from (7.5)

We are going to apply (2.9) to the function f on $\mathbf{R}^2$ defined by $f(u, v) = (u\bar{z}+v)^k \exp\big(-\pi t y^{-1}|uz+v|^2\big)$. Then (9.23) shows that $\hat{f}(r, s) = t^{-k-1}g(r, s)$ with $g(r, s) = (s\bar{z}-r)^k \exp\big(-\pi t^{-1}y^{-1}|sz-r|^2\big)$. Take x of (2.9) to be (p, q) with $p, q \in \mathbf{Z}$ and $L = N\mathbf{Z}^2$. Then (2.9) takes the form

$$\sum_{(c,\,d)\in\mathbf{Z}^2} f(p+Nc,\, q+Nd) = t^{-k-1}N^{-2}\sum_{(r,\,s)\in N^{-1}\mathbf{Z}^2} \mathbf{e}(pr+qs)g(r,\, s).$$

Let $\varphi(t)$ denote the left-hand side and $\psi(t^{-1})$ the right-hand side without the factor t^{-k-1}, suppressing the variable z. Then we have

$$\varphi(t) = t^{-k-1}\psi(t^{-1}),$$

$$\varphi(t) = \sum_{(m,n)\in(p,q)+N\mathbf{Z}^2} (m\bar{z}+n)^k \exp\bigl(-\pi t y^{-1}|mz+n|^2\bigr),$$

$$\psi(t) = N^{-k-2}\sum_{(a,b)\in\mathbf{Z}^2} \mathbf{e}\bigl((qa-pb)/N\bigr)(a\bar{z}+b)^k \exp\bigl(-\pi t y^{-1}N^{-2}|az+b|^2\bigr).$$

To obtain the expression for ψ, we put $(Ns,\, -Nr) = (a,\, b)$. Let $\varepsilon = 1$ if $k = 0$ and $\varepsilon = 0$ otherwise; let δ be as in our theorem. Put $\Phi(t) = \varphi(t) - \delta\varepsilon$ and $\Psi(t) = N^{k+2}\psi(t) - \varepsilon$. Then, by (2.1h) we have at least formally

$$\text{(1)}\qquad \int_0^\infty \Phi(t)t^{s-1}dt = \Gamma(s)\pi^{-s}y^s \sum_{0\neq(m,n)\in(p,q)+N\mathbf{Z}^2} (m\bar{z}+n)^k|mz+n|^{-2s}.$$

The last sum is absolutely convergent if $\mathrm{Re}(2s-k) > 2$ for the same reason as what we said in §9.1. Therefore by (A1.2) of the Apppendix, termwise integration is justified, and the above equality is valid for $\mathrm{Re}(2s-k) > 2$. Notice that the last sum is exactly $y^{k-s}\mathfrak{E}_k^N(z,\, s-k;\, p,\, q)$. Now $\Phi(t)$ has an estimate of type (7.3a), and so $\Phi(t) = O(e^{-ct})$ as $t \to \infty$ with a positive constant that depends only on z. The same can be said about $\Psi(t)$. For the same reason as in (1) we obtain, for $\mathrm{Re}(2s-k) > 2$,

$$\text{(2)}\qquad \int_0^\infty \Psi(t)t^{s-1}dt$$
$$= \Gamma(s)N^{2s}\pi^{-s}y^k \sum_{(a,\,b)\in\mathbf{Z}^2/N\mathbf{Z}^2} \mathbf{e}\bigl((qa-pb)/N\bigr)\mathfrak{E}_k^N(z,\, s-k;\, a,\, b).$$

Now we can apply the arguments in the proof of Theorem 3.2 to the present setting by taking N there to be 1. We have

$$\Phi(t^{-1}) = N^{-k-2}t^{k+1}\Psi(t) + \varepsilon N^{-k-2}t^{k+1} - \delta\varepsilon.$$

Let $R(s)$ resp. $R'(s)$ denote the right-hand side of (1) resp. (2). Then repeating the calculation in the proof of Theorem 3.2 with obvious modifications, we find that

$$R(k+1-s) = \int_1^\infty \Phi(t)t^{k-s}dt + N^{-k-2}\int_1^\infty \Psi(t)t^{s-1}dt - \frac{N^{-2}\varepsilon}{s} + \frac{\delta\varepsilon}{s-1}$$

for $\mathrm{Re}(s) < 0$, as $\varepsilon = 0$ if $k > 0$. The last two integrals over $[1,\, \infty)$ are convergent for every s, and so $R(s)$ is meromorphic on the whole s-plane We also find that $R(k+1-s) = N^{-k-2}R'(s)$. Substituting $s+k$ for s and arranging various factors suitably, we obtain the results as stated in our theorem.

10. Polynomial relations between Eisenstein series

10.1. For a congruence subgroup Γ of $\Gamma(1)$ we put

$$\mathfrak{M}(\Gamma) = \sum_{k=0}^{\infty} \mathscr{M}_k(\Gamma), \tag{10.1}$$

where the sum is considered within the ring of all holomorphic functions on H. Suppose we find two modular forms f and g such that $\mathfrak{M}(\Gamma) = \mathbf{C}[f, g]$, though in general we may need more than two generators. For the reason explained in the introduction, we ask the question of finding the explicit form of the expression for a given element of $\mathfrak{M}(\Gamma)$ as a polynomial in f and g. There is no clear-cut answer in the general case. If $\Gamma = \Gamma(1)$, we take $\{f, g\}$ to be $\{E_4, E_6\}$. In this case the answer can be given at least for E_{2m} with any $m > 1$ by a recurrence formula obtained from the expansion of the Weierstrass $\wp$-function, as we will show below. There is one more case in which we can find a similar recurrence formula. To state such formulas, we first recall some elementary facts on $\wp$. (In Section A4 of the Appendix we present an easy treatment of this topic and give the proof of various formulas such as (10.3b) and (10.3d).)

Given two complex numbers ω_1 and ω_2 such that $\mathrm{Im}(\omega_1/\omega_2) > 0$, we put

$$L = \mathbf{Z}\omega_1 + \mathbf{Z}\omega_2, \quad G_k = G_k(\omega_1, \omega_2) = \sum_{0 \neq \omega \in L} \omega^{-k} \qquad (4 \leq k \in 2\mathbf{Z}). \tag{10.2}$$

Then we have, as usual,

$$\wp(u) = \wp(u;\, \omega_1, \omega_2) = u^{-2} + \sum_{0 \neq \omega \in L} \{(u-\omega)^{-2} - \omega^{-2}\}, \tag{10.3a}$$

$$\wp'(u)^2 = 4\wp(u)^3 - g_2\wp(u) - g_3, \tag{10.3b}$$

$$g_2 = g_2(\omega_1, \omega_2) = 60G_4, \quad g_3 = g_3(\omega_1, \omega_2) = 140G_6, \tag{10.3c}$$

$$\wp(u) = u^{-2} + \sum_{m=2}^{\infty} (2m-1)G_{2m}u^{2m-2}, \tag{10.3d}$$

where ω_1 and ω_2 are suppresed, and $\wp' = \partial\wp/\partial u$. We often specialize (ω_1, ω_2) to $(z, 1)$ with $z \in H$. For example, we put

$$\mathbf{g}_2(z) = (2\pi i)^{-4} g_2(z, 1), \quad \mathbf{g}_3(z) = (2\pi i)^{-6} g_3(z, 1). \tag{10.3e}$$

Notice that

$$G_k(z, 1) = \mathfrak{E}_k^1(z;\, 0, 0) = 2 \cdot (2\pi i)^k \Gamma(k)^{-1} E_k(z), \tag{10.4a}$$

$$\mathbf{g}_2(z) = 20E_4(z), \qquad \mathbf{g}_3(z) = (7/3)E_6(z). \tag{10.4b}$$

We define $\Delta(z)$ and $J(z)$ as usual by

$$\Delta(z) = \mathbf{g}_2(z)^3 - 27\mathbf{g}_3(z)^2 = 20^3 E_4(z)^3 - 3 \cdot 7^2 E_6(z)^2, \tag{10.4c}$$

$$J(z) = \mathbf{g}_2(z)^3/\Delta(z). \tag{10.4d}$$

It is well known that Δ of (10.4c) coincides with Δ of (1.12), and $\mathscr{A}_0(\Gamma(1)) = \mathbf{C}(J)$. We will be considering the elliptic curve

$$(10.4e) \qquad C(\omega_1,\,\omega_2) \,:\, Y^2 = 4X^3 - g_2(\omega_1,\,\omega_2)X - g_3(\omega_1,\,\omega_2),$$

which we view as a projective curve in an obvious way. This is parametrized by the map $u \mapsto (\wp(u;\,\omega_1,\,\omega_2),\,\wp'(u;\,\omega_1,\,\omega_2))$, and isomorphic to $\mathbf{C}/(\mathbf{Z}z + \mathbf{Z})$ with $z = \omega_1/\omega_2$.

10.2. Differentiating equality (10.3b) and dividing by $2\wp'$, we obtain

$$(10.5) \qquad \wp'' = \partial^2\wp/\partial u^2 = 6\wp^2 - 2^{-1}g_2.$$

Put $\wp^{(\nu)}(u;\,\omega_1,\,\omega_2) = (\partial/\partial u)^\nu \wp(u;\,\omega_1,\,\omega_2)$. Then

$$(10.6a) \quad \wp^{(\nu)}(\lambda u;\,\lambda\omega_1,\,\lambda\omega_2) = \lambda^{-\nu-2}\wp^{(\nu)}(u;\,\omega_1,\,\omega_2) \qquad (\nu \geq 0,\ \lambda \in \mathbf{C}^\times),$$

$$(10.6b) \qquad \wp^{(\nu)}(u;\,\omega_1,\,\omega_2) = (-1)^\nu(\nu+1)!\sum_{\omega\in L}(u-\omega)^{-\nu-2} \qquad (\nu > 0).$$

Put $\wp(u) = u^{-2} + \sum_{n=2}^{\infty} c_n u^{2n-2}$. Then

$$(10.6c) \qquad \wp''(u) = 6u^{-4} + \sum_{n=2}^{\infty}(2n-2)(2n-3)c_n u^{2n-4}.$$

Substituting these into (10.5) and putting the result in order, we obtain

$$(10.7) \qquad (n-3)(2n+1)c_n = 3\sum_{r=2}^{n-2} c_r c_{n-r} \qquad (n \geq 4).$$

Since $c_n = 2(2\pi i)^{2n}\Gamma(2n-1)^{-1}E_{2n}(z)$, (10.7) gives

$$(10.8) \qquad \frac{(n-3)(2n+1)E_{2n}}{(2n-2)!} = 6\sum_{r+s=n}\frac{E_{2r}E_{2s}}{(2r-2)!(2s-2)!} \qquad (n \geq 4),$$

where $(r,\,s)$ runs over the ordered pairs of integers ≥ 2 such that $r + s = n$. This is the recurrence formula mentioned at the beginning. Relation (10.7) was known to Weierstrass; see [Wei, p. 11, (8)]. It has been reproduced in many textbooks on elliptic functions, but apparently it has never been viewed as a relation between Eisenstein series as stated here. By means of (10.8) we can find inductively an expression for E_{2n} as a polynomial in E_4 and E_6. We note the first four such expressions:

$$(10.9) \quad E_8 = 120E_4^2, \quad 11E_{10} = 5040E_4E_6, \quad 13E_{12} = 12600(96E_4^3 + E_6^2),$$
$$E_{14} = 2^8 3^4 5^2 7 \cdot E_4^2 E_6.$$

The numerical constant of the last formula can be obtained from (9.8a). We note that $\mathscr{S}_k(\Gamma(1)) = \Delta\mathscr{M}_{k-12}(\Gamma(1))$; see [S71, Proposition 2.27]. We have also $\mathscr{M}_k(\Gamma(1)) = \mathscr{S}_k(\Gamma(1)) + \mathbf{C}E_k$. Thus $\mathscr{M}_k(\Gamma(1))$ has a $\mathbf{C}$-basis formed by $\Delta^a E_b$ with a and b such that $k = 12a + b$.

10.3. We are going to state a recurrence formula for Eisenstein series of a higher level. Given $k \in \mathbf{Z}, > 0$, and $(a, b) \in \mathbf{Q}^2$, take $N \in \mathbf{Z}, > 0$, so that $N(a, b) \in \mathbf{Z}^2$. We then put $(p, q) = (Na, Nb)$, and

$$f^{(\nu)}(z;\, a,\, b) = (2\pi i)^{-\nu-2}\wp^{(\nu)}(az+b;\, z,\, 1) \qquad ((a,\, b) \notin \mathbf{Z}^2), \tag{10.10a}$$

$$F_k(z) = F_k(z;\, a,\, b) = (-2\pi i)^{-k}N^k\Gamma(k)\mathfrak{E}_k^N(z;\, p,\, q) \qquad (k \neq 2), \tag{10.10b}$$

$$F_2(z) = F_2(z;\, a,\, b) = f(z;\, a,\, b) \qquad ((a,\, b) \notin \mathbf{Z}^2), \tag{10.10c}$$

$$\mathscr{F}_2(z) = \mathscr{F}_2(z;\, a,\, b) = (2\pi i)^{-2}N^2\mathfrak{E}_2^N(z;\, p,\, q), \tag{10.10d}$$

where $f(z;\, a,\, b) = f^{(0)}(z;\, a,\, b)$. We easily see that the functions of (10.10b) and (10.10d) are well defined independently of the choice of N. These functions, except $\mathscr{F}_2$, are modular forms of level N, and $\mathscr{F}_2 \in \mathscr{N}_2^1$. They depend only on (a, b) modulo $\mathbf{Z}^2$; see (9.3d). From (10.6b) and (9.7) we easily see that

$$f^{(k-2)}(z;\, a,\, b) = F_k(z;\, a,\, b) \qquad (k \geq 2,\, (a,\, b) \notin \mathbf{Z}^2), \tag{10.11a}$$

$$F_k(z;\, 0,\, 0) = 2E_k(z) \qquad (k \neq 2), \tag{10.11b}$$

$$\mathscr{F}_2(z;\, 0,\, 0) = 2E_2(z). \tag{10.11c}$$

The Fourier expansion of $f(z;\, r/N,\, s/N)$ for $0 \leq r < N$, $(r,\, s) \notin N\mathbf{Z}^2$, is as follows:

$$\begin{aligned} f(z;\, r/N,\, s/N) = {} & \frac{1}{12} - 2\sum_{n=1}^{\infty}\Big\{\sum_{0<d|n} d\Big\}\mathbf{q}^n \\ & + \zeta^s\mathbf{q}_N^r(1-\zeta^s\mathbf{q}_N^r)^{-2} + \sum_{n=1}^{\infty} n(\zeta^{sn}\mathbf{q}_N^{rn} + \zeta^{-sn}\mathbf{q}_N^{-rn}) \cdot \sum_{m=1}^{\infty}\mathbf{q}^{mn}, \end{aligned} \tag{10.12}$$

where $\mathbf{q} = \mathbf{e}(z)$, $\mathbf{q}_N = \mathbf{e}(z/N)$, and $\zeta = \mathbf{e}(1/N)$. This was given in [S71, pp. 140–141]. For the reader's convenience, we give a proof in §A4.2 of the Appendix.

Comparing (10.12) with (9.4), we easily see that

$$F_2(z;\, a,\, b) = \mathscr{F}_2(z;\, a,\, b) - 2E_2(z). \tag{10.13}$$

This was noted by Hecke in [He27, (14)], where he said that the equality would follow from the limit process $s \mapsto 0$ applied to $\mathfrak{E}_2^N(z,\, s;\, p,\, q)$ (without employing the Fourier expansions), but this does not seem to be so easy as he might have thought.

10.4. Successive differentiation of (10.5) produces

$$\wp^{(r+3)} = 12\sum_{a=0}^{r}\binom{r}{a}\wp^{(a+1)}\wp^{(r-a)} \qquad (r \geq 0). \tag{10.14}$$

Thus, from (10.3b), (10.4b), (10.5), and (10.14) we obtain

$$F_3^2 = 4F_2^3 - 20E_4F_2 - (7/3)E_6, \tag{10.15a}$$

$$F_4 = 6F_2^2 - 10E_4, \tag{10.15b}$$

$$F_{r+5} = 12\sum_{i=0}^{r}\binom{r}{i}F_{i+3}F_{r-i+2} \qquad (r \geq 0), \tag{10.15c}$$

where F_k means $F_k(z;\, a,\, b)$ with $(a,\, b) \notin \mathbf{Z}^2$.

Theorem 10.5. *For any fixed* $(a,\, b) \in \mathbf{Q}^2$, $\notin \mathbf{Z}^2$, *and* $k \geq 2$ *the function* $F_k(z;\, a,\, b)$ *is a polynomial of* $F_2(z;\, a,\, b)$, $F_3(z;\, a,\, b)$, *and* $F_4(z;\, a,\, b)$ *with rational coefficients; it is also a polynomial of* $F_2(z;\, a,\, b)$, $F_3(z;\, a,\, b)$, *and* $E_4(z)$ *with rational coefficients.*

This follows immediately from (10.15b, c) by induction on k. For example, we have

$$F_5 = 12F_3F_2, \quad F_6 = 12F_3^2 + 12F_4F_2, \quad F_7 = 36F_4F_3 + 144F_3F_2^2. \tag{10.16}$$

Applying D_{r+5}^n to (10.15c) and using (8.9), we obtain

$$D_{r+5}^n F_{r+5} = 12\sum_{i=0}^{r}\binom{r}{i}\sum_{j=0}^{n}\binom{n}{j}(D_{i+3}^j F_{i+3})(D_{r-i+2}^{n-j}F_{r-i+2}) \tag{10.17}$$

for every $n \geq 0$. The significance of this formula will become clearer in the next section. Interestingly, these relations hold for every $(a,\, b) \in \mathbf{Q}^2$, $\notin \mathbf{Z}^2$, though we are suppressing those parameters. Also, (10.15c) and (10.17) involve modular forms of both even and odd weights, while (10.8) concerns only those of even weights.

10.6. The values of $\wp$ at the half periods produce modular forms of weight 2 and level 2. They are useful in various ways, but are not fully discussed in any modern textbook. Therefore we present here a detailed exposition of this subject. Thus we put $e_\nu = \wp(\omega_\nu/2;\, \omega_1,\, \omega_2)$ for $\nu = 1,\, 2,\, 3$, where $\omega_3 = \omega_1 + \omega_2$. It is well known that

$$4x^3 - g_2x - g_3 = 4(x - e_1)(x - e_2)(x - e_3). \tag{10.18}$$

Though these quantities are simply written e_ν and g_μ, it should be remembered that they are functions of $(\omega_1,\, \omega_2)$. We have clearly

$$e_1 + e_2 + e_3 = 0, \quad 4(e_2e_3 + e_3e_1 + e_1e_2) = -g_2, \quad 4e_1e_2e_3 = g_3. \tag{10.19a}$$

By a direct calculation we can easily verify that

$$g_2^3 - 27g_3^2 = 16(e_2 - e_3)^2(e_3 - e_1)^2(e_1 - e_2)^2. \tag{10.19b}$$

Though ω_ν is the period of $d\wp/\wp'$ over a 1-cycle, it is also practical to express $\omega_\nu/2$ as an elliptic integral. Indeed, we have clearly

$$\omega_\nu/2 = \int_{\omega_\nu/2}^{\omega_\nu} du = \int_{e_\nu}^{\infty}\frac{dx}{\sqrt{4x^3 - g_2x - g_3}}. \tag{10.20}$$

In order to view the e_ν as modular forms, we put

(10.21) $\varepsilon_1(z) = f(z;\, 1/2,\, 0),\quad \varepsilon_2(z) = f(z;\, 0,\, 1/2),\quad \varepsilon_3(z) = f(z;\, 1/2,\, 1/2).$

We have $e_\nu = (2\pi i/\omega_2)^2\varepsilon_\nu(z)$, and so

(10.22a) $$4x^3 - \mathbf{g}_2 x - \mathbf{g}_3 = 4x^3 - 20E_4 x - (7/3)E_6 = 4(x - \varepsilon_1)(x - \varepsilon_2)(x - \varepsilon_3),$$

(10.22b) $$\varepsilon_1 + \varepsilon_2 + \varepsilon_3 = 0,\quad 4(\varepsilon_2\varepsilon_3 + \varepsilon_3\varepsilon_1 + \varepsilon_1\varepsilon_2) = -\mathbf{g}_2,\quad 4\varepsilon_1\varepsilon_2\varepsilon_3 = \mathbf{g}_3,$$

(10.22c) $$\Delta = 16(\varepsilon_2 - \varepsilon_3)^2(\varepsilon_3 - \varepsilon_1)^2(\varepsilon_1 - \varepsilon_2)^2.$$

The variable z is suppressed in these formulas. We note here well known formulas:

(10.23a) $$\varepsilon_1(z+1) = \varepsilon_3(z),\quad \varepsilon_2(z+1) = \varepsilon_2(z),\quad \varepsilon_3(z+1) = \varepsilon_1(z),$$

(10.23b) $$\varepsilon_1(-z^{-1}) = z^2\varepsilon_2(z),\quad \varepsilon_2(-z^{-1}) = z^2\varepsilon_1(z),\quad \varepsilon_3(-z^{-1}) = z^2\varepsilon_3(z),$$

(10.23c) $$\varepsilon_2(z) = -2\varepsilon_1(2z).$$

Clearly (10.23a) and (10.23b) follow from (10.11a), (10.13), and (9.3). As for (10.23c), we note that $\varepsilon_2 \in \mathscr{M}_2(\Gamma_0(2))$ and $\varepsilon_1 \in \mathscr{M}_2(\alpha\Gamma_0(2)\alpha^{-1})$ with $\alpha = \mathrm{diag}[2,\, 1]$; also, $\mathscr{M}_2(\Gamma_0(2))$ is of dimension 1, as can easily be seen from the standard fact on such a dimension; see [S71, Theorem 2.23]. Therefore, comparing the constant terms of these functions, we obtain (10.23c).

The Fourier expansions of the functions ε_ν are as follows:

(10.24a) $$\varepsilon_1(z) = \frac{1}{12} + 2\sum_{n=1}^{\infty}\Bigl\{\sum_{d\ \mathrm{odd},\ >0,\ d|n} d\Bigr\}\mathbf{q}_2^n,$$

(10.24b) $$\varepsilon_2(z) = \frac{-1}{6} - 4\sum_{n=1}^{\infty}\Bigl\{\sum_{d\ \mathrm{odd},\ >0,\ d|n} d\Bigr\}\mathbf{q}^n,$$

(10.24c) $$\varepsilon_3(z) = \frac{1}{12} + 2\sum_{n=1}^{\infty}(-1)^n\Bigl\{\sum_{d\ \mathrm{odd},\ >0,\ d|n} d\Bigr\}\mathbf{q}_2^n,$$

where $\mathbf{q}_2 = \mathbf{e}(z/2)$. Indeed, we obtain (10.24b) immediately from (10.12). Then (10.24a) and (10.24c) follow from (10.23c) and (10.23a).

We note interesting equalities

(10.24d) $$-2\pi\varepsilon_1(z) = \mathfrak{E}_2^2(z,\, -1;\, 1,\, 0),\quad -2\pi\varepsilon_2(z) = \mathfrak{E}_2^2(z,\, -1;\, 0,\, 1),$$
$$-2\pi\varepsilon_3(z) = \mathfrak{E}_2^2(z,\, -1;\, 1,\, 1).$$

The first equality can be shown by comparing (10.24a) with (9.14). Then the transformations $z \mapsto -z^{-1}$ and $z \mapsto z+1$ prove the other two.

10.7. As explained in §10.2, for $\Gamma = SL_2(\mathbf{Z})$ the question of expressing an Eisenstein series as a polynomial in E_4 and E_6 is settled by (10.8). The problem for a more general Γ is not so simple. Let us now treat relatively easy types of Γ of level 4. We first observe that $F_k(z;\, 0,\, 1/4)$ belongs to $\mathscr{M}_k(\Gamma_1(4))$. Put $\Gamma_\nu =$

$\gamma_\nu^{-1}\Gamma_1(4)\gamma_\nu$ with $\gamma_1 = \begin{bmatrix} 0 & -1 \\ 1 & 0 \end{bmatrix}$, $\gamma_2 = 1$, and $\gamma_3 = \begin{bmatrix} 1 & 0 \\ 1 & 1 \end{bmatrix}$; also put $(a_1, b_1) = (1/4, 0)$, $(a_2, b_2) = (0, 1/4)$, $(a_3, b_3) = (1/4, 1/4)$. Then $(0, 1/4)\gamma_\nu = (a_\nu, b_\nu)$, and so $F_k(z; a_\nu, b_\nu)$ belongs to $\mathscr{M}_k(\Gamma_\nu)$ for $k \geq 1$; in addition, $\varepsilon_\nu \in \mathscr{M}_2(\Gamma_\nu)$. Fixing ν, let us simply denote by F_k and $\mathscr{F}_2$ the functions defined by (10.10b, c, d) with $(a, b) = (a_\nu, b_\nu)$. Then we have

(10.25a) $$\dim \mathscr{M}_k(\Gamma_\nu) = 1 + [k/2],$$

(10.25b) $$\mathfrak{M}(\Gamma_\nu) = \mathbf{C}[F_1, \varepsilon_\nu] \qquad (\nu=1, 2, 3).$$

It is sufficient to prove these for one particular choice of ν, say $\nu = 2$. Then $\Gamma_2 = \Gamma_1(4)$. For that group we have proved (10.25a) in (1.22).

Next, from (9.4) we see that

$$F_1(z; 0, 1/4) = 2^{-1}i + 2i\sum_{n=1}^{\infty} \mathbf{e}(nz)\Big\{\sum_{0<d|n}\Big(\frac{-4}{d}\Big)\Big\},$$

which equals $2i$ times $\varphi_{1,\chi}^{\tau}$ of (9.10b) with $N = 4$. The Fourier expansion of ε_2 is given by (10.24b), and we easily see that F_1^2 and ε_2 are linearly independent over $\mathbf{C}$. Suppose F_1 and ε_2 are algebraically dependent over $\mathbf{C}$. Then, since modular forms of different weights are linearly independent over $\mathbf{C}$, there is a nontrivial relation $\sum_{a=0}^{m} c_a F_1^{n-2a}\varepsilon_2^a = 0$ with $0 < m \leq n/2$ and $c_a \in \mathbf{C}$. This means that $F_1^{-2}\varepsilon_2$ is a constant, a contradiction. Now the number of pairs (a, b) of nonnegative integers a, b such that $a + 2b = k$ is exactly $1 + [k/2]$. The monomials $F_1^a\varepsilon_2^b$ for such (a, b) are linearly independent over $\mathbf{C}$, and so they span $\mathscr{M}_k(\Gamma)$. Thus we obtain (10.25b).

Now we have

(10.26a) $$2F_2 = 2\mathscr{F}_2 - 4E_2 = 4F_1^2 - \varepsilon_\nu,$$

(10.26b) $$F_3 = 8F_1^3 - 6F_1\varepsilon_\nu,$$

(10.26c) $$F_4 = 6F_2^2 - 10E_4,$$

(10.26d) $$5E_4 = -4F_1^4 + 6F_1^2\varepsilon_\nu + (3/4)\varepsilon_\nu^2,$$

(10.26e) $$7E_6 = 12\varepsilon_\nu^3 - 60\varepsilon_\nu E_4,$$

(10.26f) $$D_1F_1 + 2F_1E_2 = 2F_1^3 - (5/2)F_1\varepsilon_\nu,$$

(10.26g) $$D_2\varepsilon_\nu + 4E_2\varepsilon_\nu = 2\varepsilon_\nu^2 - (20/3)E_4,$$

(10.26h) $$\mathfrak{M}(\Gamma_\nu) = \mathbf{C}[F_1, F_2].$$

The last equality follows from (10.25b) and (10.26a). We already noted (10.26c) in (10.15b); (10.226e) follows from the fact that ε_ν is a root of the polynomial of (10.22a); the remaining equalities can be verified by comparing the first few Fourier coefficients of both sides. As for F_k with $k \geq 5$, we can use (10.15c). These together with (9.9a, b, c, d), (10.15b), and (10.17) are enough for concluding that $D_k^n F_k$ and $D_2^n\mathscr{F}_2$ are polynomials in F_1, ε_ν, and E_2 with rational coefficients which are computable.

Put now

$$F_k^{(N)}(z;\, p/N,\, q/N) = (-2i)^{-k} N\pi^{-1} \mathfrak{E}_k^N(z,\, 1-k;\, p,\, q) \quad (1<k\in\mathbf{Z}). \tag{10.27}$$

The Fourier expansion of this function was given in (9.14). This function $F_k^{(N)}$ is often useful. Notice that it depends on N. If $N = 1$, this coincides with $2E_k(z)$ by (9.22). For $N = 4$ and $(a,\, b) = (a_\nu,\, b_\nu)$, $\nu = 1, 2, 3$, as above we have

$$F_3^{(4)}(z;\, a,\, b) = 4F_3(z;\, a,\, b) - 16F_1(z;\, a,\, b)^3. \tag{10.27a}$$

In (9.17a, b) we stated formulas for $\mathfrak{E}_k^N(z,\, -n;\, p,\, q)$. Here we reformulate them in terms of $F_k^*(z;\, p/N,\, q/N)$ and $F_k^{(N)}(z;\, p/N,\, q/N)$ as follows:

$$\mathfrak{E}_\nu^N(z,\, -n;\, p,\, q) = \pi^{(\nu+k)/2}\Gamma\big((\nu+k)/2\big)^{-1}(-2i)^\nu \cdot \begin{cases} N^{-k} D_k^n F_k^*(z;\, p/N,\, q/N) & (k>0), \\ N^{-1} D_{2-k}^m F_{2-k}^{(N)}(z;\, p/N,\, q/N) & (k\le 0). \end{cases} \tag{10.28}$$

Here $\nu,\, k \in \mathbf{Z}$, $2-\nu \le k \le \nu$, $n = (\nu-k)/2 \in \mathbf{Z}$, and $m = (\nu+k-2)/2$; $F_k^* = F_k$ if $k \ne 2$ and $F_2^* = \mathscr{F}_2$. Thus we need $D_{2-k}^m F_{2-k}^{(N)}$ in addition to $D_k^n F_k^*$ and we naturally ask whether $D_{2-k}^m F_{2-k}^{(N)}$ is a polynomial in the basic functions such as F_1, ε_ν, and E_2 for Γ_ν of level 4. We will answer this affirmatively in Lemma 10.14 below at least when the group is conjugate to $\Gamma_1(N)$ with $3 \le N \le 6$.

10.8. Let us next discuss the case of level 3. Let $\Gamma = \gamma^{-1}\Gamma_1(3)\gamma$ with a fixed $\gamma \in SL_2(\mathbf{Z})$. Put $F_k = F_k(z;\, c_\gamma/3,\, d_\gamma/3)$ for simplicity. Then by the same technique as in §10.7 we can show that $\dim \mathscr{M}_k(\Gamma) = 1 + [k/3]$ and $\mathfrak{M}(\Gamma) = \mathbf{C}[F_1,\, F_3]$. The function $F_k(z;\, 1/3,\, 0)$ for odd k is the same as the function of (9.10a) for $N = 3$. Using the Fourier expansions, we can verify that

$$F_2 = \mathscr{F}_2 - 2E_2 = 3F_1^2, \tag{10.29a}$$

$$5E_4 = 27F_1^4 - 3F_1F_3, \tag{10.29b}$$

$$F_4 = 6F_1F_3, \tag{10.29c}$$

$$7E_6 = -3F_3^2 + 2^2\cdot 3^3 F_1^3 F_3 - 2^3\cdot 3^4 F_1^6, \tag{10.29d}$$

$$D_1F_1 + 2F_1E_2 = 2^{-1}F_3 - 3F_1^3, \tag{10.29e}$$

$$D_3F_3 + 6F_3E_2 = 3F_1^2F_3. \tag{10.29f}$$

Notice that (10.29c) follows from (10.15b) and (10.29a, b). For the same reason as in §10.7, $D_k^n F_k$ and $D_2^n \mathscr{F}_2$ are polynomials in F_1, F_3, and E_2 with rational coefficients which are computable.

Using the symbol of (10.27), for $(a,\, b) \in 3^{-1}\mathbf{Z}^2$, $\notin \mathbf{Z}^2$ we have

$$F_3^{(3)}(z;\, a,\, b) = 3F_3(z;\, a,\, b) - 24F_1(z;\, a,\, b)^3, \tag{10.30a}$$

$$F_4^{(3)}(z;\, a,\, b) = F_4(z;\, a,\, b) - 26E_4(z). \tag{10.30b}$$

10.9. Let us now recall some basic properties of the classical modular function λ which is defined by

$$\lambda(z) = \frac{\varepsilon_3(z) - \varepsilon_1(z)}{\varepsilon_2(z) - \varepsilon_1(z)} = \frac{e_3 - e_1}{e_2 - e_1} \qquad (z \in H), \tag{10.31}$$

where ε_ν and e_ν are as in §10.5. We see that $\lambda \in \mathscr{A}_0(\Gamma(2))$, as $\varepsilon_\nu \in \mathscr{M}_2(\Gamma(2))$. The e_ν are all different, and so $\lambda(z) \neq 0, \infty$ for every $z \in H$.

Theorem 10.10. (i) $\mathscr{A}_0(\Gamma(2)) = \mathbf{C}(\lambda)$.
(ii) $\mathscr{A}_0(\Gamma_0(2)) = \mathbf{C}(\mu)$, *where* $\mu = (1-\lambda)\lambda^{-2}$.

(iii) $J = \dfrac{4(\mu+1)^3}{27\mu^2} = \dfrac{4(\lambda^2-\lambda+1)^3}{27\lambda^2(1-\lambda)^2}$.

(iv) *The function* λ *maps* $\Gamma(2)\backslash H$ *bijectively onto* $\mathbf{C} - \{0, 1, \infty\}$.

PROOF. We have

$$\mu = (\varepsilon_2 - \varepsilon_1)(\varepsilon_2 - \varepsilon_3)(\varepsilon_3 - \varepsilon_1)^{-2}. \tag{10.32}$$

From (10.23a) we see that $\mu(z+1) = \mu(z)$, and so $\mu \in \mathscr{A}_0(\Gamma_0(2))$, as $\Gamma_0(2)$ is generated by $\Gamma(2)$ and $\begin{bmatrix} 1 & 1 \\ 0 & 1 \end{bmatrix}$. Since $\varepsilon_1 + \varepsilon_2 + \varepsilon_3 = 0$, we can verify that $\mu + 1 = -3(\varepsilon_2\varepsilon_3 + \varepsilon_3\varepsilon_1 + \varepsilon_1\varepsilon_2)(\varepsilon_3 - \varepsilon_1)^{-2}$, and so by (10.22b, c),

$$\frac{4(\mu+1)^3}{27\mu^2} = \frac{-4(\varepsilon_2\varepsilon_3 + \varepsilon_3\varepsilon_1 + \varepsilon_1\varepsilon_2)^3}{(\varepsilon_2 - \varepsilon_3)^2(\varepsilon_3 - \varepsilon_1)^2(\varepsilon_1 - \varepsilon_2)^2} = \frac{\mathbf{g}_2^3}{\Delta} = J.$$

This proves the first equality of (iii). Then the second part can easily be verified. Now the equation $27\mu^2 J = 4(\mu+1)^3$ shows that $[\mathbf{C}(\mu) : \mathbf{C}(J)] = 3$. Since $\mu \in \mathscr{A}_0(\Gamma_0(2))$ and $[\Gamma(1) : \Gamma_0(2)] = 3$, we obtain (ii), and also (i), as $[\mathbf{C}(\lambda) : \mathbf{C}(J)] = 6$ and $[\Gamma(1) : \Gamma(2)] = 6$. In view of (i), we see that λ gives a bijection of the compact Riemann surface $\Gamma(2)\backslash(H \cup \mathbf{Q} \cup \infty)$ onto the Riemann sphere $\mathbf{C} \cup \{\infty\}$. We have $J(z) \neq \infty$ for $z \in H$, and so the formula of (iii) shows that $\lambda(z) \notin \{0, 1, \infty\}$. Since we can easily show that $\Gamma(2)\backslash(\mathbf{Q} \cup \infty)$ has exactly 3 orbits, we obtain (iv).

Lemma 10.11. *Let* $x = \varepsilon_1/\varepsilon_2$ *and* $w = \mathbf{g}_2\varepsilon_2^{-2}$. *Then* $w = 4(x^2+x+1)$, $4\mu = (12-w)/(w-3)$, *and*

$$J = w^3(w-3)^{-1}(w-12)^{-2}. \tag{10.33}$$

PROOF. Since $\varepsilon_3 = -\varepsilon_1 - \varepsilon_2$, from (10.32) we obtain

$$\mu = \frac{(\varepsilon_2 - \varepsilon_1)(2\varepsilon_2 + \varepsilon_1)}{(2\varepsilon_1 + \varepsilon_2)^2} = \frac{(1-x)(2+x)}{(2x+1)^2} = \frac{2-x-x^2}{4x^2+4x+1}.$$

On the other hand, by (10.22b) we have $w = -4(\varepsilon_2\varepsilon_3 + \varepsilon_3\varepsilon_1 + \varepsilon_1\varepsilon_2)\varepsilon_2^{-2} = 4(\varepsilon_1^2+\varepsilon_1\varepsilon_2+\varepsilon_2^2)\varepsilon_2^{-2} = 4(x^2+x+1)$, and so $-4\mu = (w-12)/(w-3)$. Substituting this into the equality $J = 4(\mu+1)^3/(27\mu^2)$, we obtain (10.33).

Notice that (10.33) holds for $w = 20E_4\varepsilon_\nu^{-2}$, $\nu = 1, 2, 3$, as $\Gamma(1)$ permutes the ε_ν. Also, the equality $4\mu = (12-w)/(w-3)$ shows that $\mathscr{A}_0(\Gamma_0(2)) = \mathbf{C}(\mathbf{g}_2\varepsilon_2^{-2})$.

In the following theorem, for a subfield R of $\mathbf{C}$ and functions $f_1, \ldots, f_\mu$, the symbol $R[f_1, \ldots, f_\mu]$ denotes the ring generated by R and $f_1, \ldots, f_\mu$.

Theorem 10.12. *Let Γ be a congruence subgroup of $SL_2(\mathbf{Z})$; suppose that $\mathfrak{M}(\Gamma) = \mathbf{C}[f_1, \ldots, f_\mu]$ with $f_\nu \in \mathscr{M}_{\kappa_\nu}(\mathbf{Q})$. Then*

$$\sum_{k=0}^{\infty} \mathscr{M}_k(\mathbf{Q}, \Gamma) = \mathbf{Q}[f_1, \ldots, f_\mu].$$

PROOF. Put $\varphi_a = f_1^{a_1} \cdots f_\mu^{a_\mu}$ for $a = (a_1, \ldots, a_\mu)$ with nonnegative integers a_i. Then we can find a $\mathbf{C}$-basis of $\mathfrak{M}(\Gamma)$ consisting of such φ_a. Thus, given $h \in \mathscr{M}_k(\mathbf{Q}, \Gamma)$ with any k, we can put $h = \sum_a c_a\varphi_a$ with $c_a \in \mathbf{C}$. Clearly $c_a \neq 0$ only when $k = a_1\kappa_1 + \cdots + a_\mu\kappa_\mu$. By Theorem 1.9, all such φ_a and h cannot be linearly independent over $\mathbf{Q}$. Therefore $h \in \mathbf{Q}[f_1, \ldots, f_\mu]$ as expected.

10.13. We apply the above theorem to the group $\Gamma = \Gamma_\alpha = \alpha^{-1}\Gamma_1(N)\alpha$ with $3 \leq N \leq 6$ and $\alpha \in \Gamma(1)$. We have $\mathfrak{M}(\Gamma_1(N)) = \mathbf{C}[\varphi, \psi]$, where

$$\varphi(z) = F_1(z;\, 0,\, 1/N), \qquad \psi(z) = \begin{cases} F_3(z;\, 0,\, 1/3) & (N=3), \\ F_2(z;\, 0,\, 1/4) & (N=4), \\ F_1(z;\, 0,\, 2/N) & (N = 5 \text{ or } 6). \end{cases}$$

We have shown this in §10.7 for $N = 3$ and in §10.8 for $N = 4$. The case $N = 5$ or 6 can be proved in the same way. Therefore $\mathfrak{M}(\Gamma_\alpha) = \mathbf{C}[\varphi_\alpha, \psi_\alpha]$, where

$$\varphi_\alpha(z) = F_1(z;\, c_\alpha/N,\, d_\alpha/N), \quad \psi_\alpha(z) = \begin{cases} F_3(z;\, c_\alpha/3,\, d_\alpha/3) & (N=3), \\ F_2(z;\, c_\alpha/4,\, d_\alpha/4) & (N=4), \\ F_1(z;\, 2c_\alpha/N,\, 2d_\alpha/N) & (N = 5 \text{ or } 6). \end{cases}$$

Lemma 10.14. *The functions $F_k(z;\, c_\alpha/N,\, d_\alpha/N)$ and $F_k^{(N)}(z;\, c_\alpha/N,\, d_\alpha/N)$ defined for N and α as above belong to $\mathbf{Q}[\varphi_\alpha, \psi_\alpha]$.*

PROOF. First take α to be $\iota = \begin{bmatrix} 0 & -1 \\ 1 & 0 \end{bmatrix}$. Then

$$F_k(z;\, c_\iota/N,\, d_\iota/N) = F_k(z;\, 1/N,\, 0).$$

We observe that this function, as well as φ_ι, ψ_ι, and $F_k^{(N)}(z;\, 1/N,\, 0)$, have rational Fourier coefficients at ∞. Thus Theorem 10.12 shows that $\sum_{k=0}^{\infty} \mathscr{M}_k(\mathbf{Q}, \Gamma_\iota) = \mathbf{Q}[\varphi_\iota, \psi_\iota]$. This proves our assertion when $\alpha = \iota$. Transforming this by $\iota^{-1}\alpha$, we obtain the desired result for an arbitrary α.

It should be noted that the functions of the above lemma do not necessarily belong to $\mathscr{M}_k(\mathbf{Q})$.

10.15. Let us now discuss a case in which $\mathfrak{M}(\Gamma)$ requires more than two generators. We take a real Dirichlet character χ of conductor 7, that is, $\chi(a) = \left(\frac{-7}{a}\right)$, and put

$$\Gamma = \{\gamma \in \Gamma_0(7) \mid \chi(d_\gamma) = 1\}. \tag{10.34}$$

Then $\Gamma_0(7) = \{\pm 1\}\Gamma$. By means of the general principles explained in §1.11 we can easily verify that for every $k \in \mathbf{Z}, > 0$,

$$\dim \mathscr{M}_k(\Gamma) = 2[k/3] + 1, \quad \dim \mathscr{S}_k(\Gamma) = \mathrm{Max}(2[k/3] - 1, 0). \tag{10.35}$$

Using the notation of (9.10a, b) with $N = 7$, we put

$$\varphi_k(z) = \varphi^r_{k,\chi}(z), \qquad \psi_k(z) = \varphi^\ell_{k,\chi}(7z) \qquad (0 \le k - 1 \in 2\mathbf{Z}). \tag{10.36}$$

By (9.11e, f) both φ_k and ψ_k belong to $\mathscr{M}_k(\Gamma)$. Let us now prove

$$\mathfrak{M}(\Gamma) = \mathbf{C}[\varphi_1, \varphi_3, \psi_3], \tag{10.37a}$$

$$\sum_{k=0}^{\infty} \mathscr{M}_k(\mathbf{Q}, \Gamma) = \mathbf{Q}[\varphi_1, \varphi_3, \psi_3]. \tag{10.37b}$$

From (9.11a, b) we see that $\varphi_k, \psi_k \in \mathscr{M}_k(\mathbf{Q})$, and so (10.37b) follows from (10.37a) and Theorem 10.12.

To prove (10.37a), we first observe that $\mathscr{M}_1(\Gamma) = \mathbf{C}\varphi_1$ and $\mathscr{M}_2(\Gamma) = \mathbf{C}\varphi_1^2$, as these spaces are one-dimensional. Next, for $k = 3\nu + a$ with $\nu > 0$ and $0 \le a \le 2$ we have $\dim \mathscr{M}_k(\Gamma) = \dim \mathscr{M}_{3\nu}(\Gamma)$, and so $\mathscr{M}_k(\Gamma) = \varphi_1^a \mathscr{M}_{3\nu}(\Gamma)$. Thus it is sufficient to show that $\mathscr{M}_{3\nu}(\Gamma) \subset \mathbf{C}[\varphi_1, \varphi_3, \psi_3]$. Now the cusps of Γ are represented by ∞ and 0. From (9.11a, b, c) we see that φ_3 is nonzero at ∞ and 0 at 0, while ψ_3 is 0 at ∞ and nonzero at 0. Therefore given $f \in \mathscr{M}_{3\nu}(\Gamma)$, we can find $a, b \in \mathbf{C}$ such that $f - a\varphi_3^\nu - b\psi_3^\nu \in \mathscr{S}_{3\nu}(\Gamma)$. Thus our task is to show that $\mathscr{S}_{3\nu}(\Gamma) \subset \mathbf{C}[\varphi_1, \varphi_3, \psi_3]$, which we do by induction on ν. We have $\dim \mathscr{S}_3(\Gamma) = 1$, and if we put

$$h = (32/3)\varphi_1^3 + (7/6)\varphi_3 - (49/6)\psi_3, \tag{10.38}$$

we easily find that this is a cusp form; in fact, $h(z) = \left[\Delta(z)\Delta(7z)\right]^{1/8}$, as can easily be verified. Therefore $\mathscr{S}_3(\Gamma) = \mathbf{C}h$. Since $\dim \mathscr{S}_{3\nu+3}(\Gamma) = \dim \mathscr{M}_{3\nu}(\Gamma)$, we see that $\mathscr{S}_{3\nu+3}(\Gamma) = h\mathscr{M}_{3\nu}(\Gamma)$, which gives the desired fact by induction, and the proof of (10.37a) is complete.

11. Recurrence formulas for the critical values of certain Dirichlet series

11.1. In this section we prove several recurrence formulas for the values of elementary Dirichlet series at integers. We first consider $D^r_{a,N}(s)$ of (3.11). For $0 < k \in \mathbf{Z}$ and $0 \le a < N$ define $V^k_{a,N}$ by

(11.1a) $$V_{a,N}^1 = \zeta^a/(\zeta^a - 1), \qquad \zeta = \mathbf{e}(1/N),$$

(11.1b) $$V_{a,N}^k = (2\pi i)^{-k}(k-1)!N^k \cdot \begin{cases} D_{a,N}^0(k) & (0 < k \in 2\mathbf{Z}), \\ D_{a,N}^1(k) & (0 < k-1 \in 2\mathbf{Z}). \end{cases}$$

Then we have, by (4.10a, b),

(11.2) $$V_{a,N}^k = \zeta^a P_k(\zeta^a)(\zeta^a - 1)^{-k} \qquad (0 < k \in \mathbf{Z}),$$

where we understand that $P_1 = 1$. By (2.20) we find that these $V_{a,N}^k$ satisfy a recurrence formula

(11.3) $$V_{a,N}^{n+1} = (\zeta^a - 1)^{-1} \sum_{k=0}^{n-1} \binom{n}{k} V_{a,N}^{k+1}.$$

Similarly, substituting a/N for t in (4.8b) and employing (4.13), we can state a recurrence formula for $D_{a,N}^r(r+1-2m)$. Also we remind the reader of the recurrence formulas for $L(k, \chi)$ given in Theorem 4.15.

Now the constant term of $F_k(z;\, 0,\, q/N)$ is $(-1)^k V_{q,N}^k$ if $k > 2$, and $12^{-1} + V_{q,N}^2$ if $k = 2$. Therefore, putting simply

(11.4) $$C_k = (-1)^k V_{q,N}^k \qquad (0 < k \in \mathbf{Z}),$$

and fixing (q, N), we obtain, from (10.15c),

(11.5) $$C_{n+5} = C_{n+3} + 12 \sum_{a=0}^{n} \binom{n}{a} C_{a+3} C_{n-a+2} \qquad (n \geq 0).$$

The value C_k can be given by (11.2), but once C_2, C_3, and C_4 are determined, C_k for larger k can easily be obtained from (11.5); for instance, $C_5 = C_3(12C_2 + 1)$, $C_6 = C_4(12C_2 + 1) + 12C_3^2$, $C_7 = C_5(12C_2 + 1) + 36C_3C_4$. Equality (11.5) shows that (2.21) holds for $x = \zeta_q = \mathbf{e}(q/N)$ whenever $0 < q < N$. Therefore we obtain (2.21).

In the easiest case with $N = 2$ and $q = 1$ we have $C_k \neq 0$ only for even k. From (4.18a) and (4.25) we see that $C_k = (1 - 2^k)B_k/k$. Thus putting

(11.6) $$A_k = (1 - 2^k)B_k/k,$$

we obtain, for $0 \leq n - 1 \in 2\mathbf{Z}$,

(11.7) $$A_{n+5} = A_{n+3} + 12 \sum_{b=1}^{(n+1)/2} \binom{n}{2b-1} A_{2b+2} A_{n+3-2b}.$$

Also, from (4.3b) and (4.40) we obtain

(11.8) $$A_{n+1} = \frac{-1}{2} \sum_{k=1}^{n} \binom{n}{k-1} A_k \qquad (0 < n \in \mathbf{Z}),$$

where $A_1 = 1/2$ and $A_k = 0$ for $0 < k - 1 \in 2\mathbf{Z}$.

11.2. Next let d denote 3 or 4, and μ_d the primitive Dirichlet character modulo d. We then consider $D^r_{1,d}(k)$ for $k-r \in 2\mathbf{Z}$. For $r = 0$ the series is given in (4.18a,b); also we easily see that $D^1_{1,d}(s) = L(s,\, \mu_d)$. Therefore C_k in this case is given by

$$(11.9)\qquad C_k = \begin{cases} (-2\pi i)^{-k} d^k \Gamma(k) L(k,\, \mu_d) & (0 < k-1 \in 2\mathbf{Z}), \\ (2\pi i)^{-k}\Gamma(k)(3^k-1)\zeta(k) & (0 < k \in 2\mathbf{Z},\ d = 3), \\ (2\pi i)^{-k}\Gamma(k)2^k(2^k-1)\zeta(k) & (0 < k \in 2\mathbf{Z},\ d = 4). \end{cases}$$

Thus (11.5) holds for these two sets of $\{C_k\}$. In particular, if $d = 4$, then C_{2n} is a rational number times B_{2n}, and C_{2n+1} is a rational number times the Euler number $E_{1,2n}$ as shown in (4.38). Therefore (11.5) in this case produces two types of recurrence formulas (depending on the parity of n) involving both Bernoulli and Euler numbers.

11.3. Comparing the constant terms on both sides of (10.8), from (9.8) and (9.8a) we obtain

$$(11.10)\qquad (n-3)(4n^2-1)\zeta(2n) = 6 \sum_{r+s=n} (2r-1)(2s-1)\zeta(2r)\zeta(2s) \qquad (n \geq 4),$$

$$(11.11)\qquad (n-3)(2n+1)\frac{|B_{2n}|}{n} = \frac{3}{2}\sum_{r+s=n} \frac{(2n-2)!}{(2r-2)!(2s-2)!} \cdot \frac{|B_{2r}|}{r}\frac{|B_{2s}|}{s} \qquad (n \geq 4).$$

Here, as well as in the following (11.12), $(r,\, s)$ runs over the ordered pairs of integers ≥ 2 such that $r + s = n$. Similarly, comparing the coefficients of the terms of (10.8) involving $\mathbf{e}(z)$, we obtain, for $n \geq 4$,

$$(11.12)\qquad \frac{(n-3)(2n+1)}{3} = \sum_{r+s=n} \frac{(2n-2)!}{(2r-2)!(2s-2)!}\left\{\frac{(-1)^r|B_{2r}|}{2r} + \frac{(-1)^s|B_{2s}|}{2s}\right\}.$$

Various recurrence formulas for B_{2n} were obtained by Ramanujan [Ram], but they do not seem to include (11.7), (11.8), (11.11), or (11.12).

CHAPTER V

CRITICAL VALUES OF DIRICHLET SERIES ASSOCIATED WITH IMAGINARY QUADRATIC FIELDS

12. The singular values of nearly holomorphic forms

12.1. Throughout the rest of the book (except in Section 15) we denote by K an imaginary quadratic field embedded in $\mathbf{C}$, by $\mathfrak{r}$ the maximal order of K, and by ρ the restriction of complex conjugation to K. For $0 > a \in \mathbf{R}$ we understand that $\sqrt{a} = i\sqrt{|a|}$. We also denote by K_{ab} the maximal abelian extension of K in $\mathbf{C}$. We first note that every element of $K \cap H$ is an elliptic fixed point of an element of $GL_2(\mathbf{Q})$. Indeed, for $s \in K$, multiplication by s on K is a $\mathbf{Q}$-linear endomorphism of K. Take $\tau \in K \cap H$ and denote by $q(s)$ the element of $M_2(\mathbf{Q})$ that represents this endomorphism with respect to the basis $\{\tau, 1\}$. Then

$$\begin{bmatrix} \tau \\ 1 \end{bmatrix} s = q(s) \begin{bmatrix} \tau \\ 1 \end{bmatrix}. \tag{12.1}$$

Clearly q gives an injection of K into $M_2(\mathbf{Q})$. If $\beta = q(s)$ with $s \in K^\times$, then $\det(\beta) = ss^\rho > 0$, and from (1.3) and (12.1) we see that $\beta(\tau) = \tau$ and $j_\beta(\tau) = s/|s|$. Moreover, we have

$$q(K^\times) = \{\alpha \in G_+ \mid \alpha(\tau) = \tau\}. \tag{12.1a}$$

To see this, we first observe that the group $\{\alpha \in G_{\mathbf{a}+} \mid \alpha(z) = z\}$ is commutative for every fixed $z \in H$, as we easily see from (1.3) that the map $\alpha \mapsto c_\alpha z + d_\alpha$ sends the group injectively into $\mathbf{C}^\times$. Now we have seen that the right-hand side of (12.1a) contains the left-hand side. Take $\alpha \in G_+$ such that $\alpha(\tau) = \tau$; take also an element s of $K^\times$ not contained in $\mathbf{Q}$ and put $\beta = q(s)$. Then $\alpha\beta = \beta\alpha$, as the right-hand side of (12.1a) is commutative. It is well known that the commutator of $q(K)$ in $M_2(\mathbf{Q})$ is $q(K)$ itself, and so $\alpha \in q(K^\times)$, which proves (12.1a).

Next, we have a basic theorem about the values of the elements of $\mathscr{N}_k(\mathbf{Q}_{\mathrm{ab}})$ at the points of $K \cap H$. Before stating it, we make a simple observation. Given $\tau \in K \cap H$, take $\alpha \in G_+$, not contained in $\mathbf{Q}1_2$, such that $\alpha(\tau) = \tau$ and put $\kappa = j_\alpha(\tau)$. Then $\kappa^2 \in K$. Indeed, if $\alpha = q(s)$, then $\kappa^2 = s/s^\rho$. Given $\varphi \in \mathscr{M}_k$ such that $\varphi(\tau) \neq 0$, put $\psi = (\varphi\|_k\alpha)/\varphi$. Then $\psi(\tau) = \kappa^{-k}$. Applying

D_k to the equality $\varphi\|_k\alpha = \varphi\psi$, we obtain, by (8.8) and (8.9), $(D_k\varphi)\|_{k+2}\alpha = \psi D_k\varphi + \varphi\cdot(2\pi i)^{-1}\psi'$, where $\psi' = d\psi/dz$. Evaluating this at $z = \tau$, we find that

$$(12.2)\qquad (\varphi^{-1}D_k\varphi)(\tau) = (\kappa^{-2} - 1)^{-1}(2\pi i)^{-1}(\psi'/\psi)(\tau).$$

Notice that $\kappa^2 \neq 1$, as $\alpha \notin \mathbf{Q}1_2$.

Now for every $h \in \mathscr{A}_0(\mathbf{Q}_{\mathrm{ab}})$ that is finite at τ, the value $h(\tau)$ belongs to K_{ab}. This is a fundamental fact of complex multiplication; see [S71, Chapter 5] and [S98]. Therefore, if $f, g \in \mathscr{M}_m(\mathbf{Q}_{\mathrm{ab}})$ and $g(\tau) \neq 0$, then $f(\tau)/g(\tau) \in K_{\mathrm{ab}}$. This can be generalized as follows.

Theorem 12.2. (i) *Let $\tau_1, \tau_2 \in K \cap H$ and $f \in \mathscr{N}_m(\mathbf{Q}_{\mathrm{ab}})$; let g be an element of $\mathscr{A}_m(\mathbf{Q}_{\mathrm{ab}})$ such that $g(\tau_2) \neq 0, \infty$. Then $f(\tau_1)/g(\tau_2) \in K_{\mathrm{ab}}$.*

(ii) *Let $\varphi \in \mathscr{N}_k(\mathbf{Q}_{\mathrm{ab}})$ and $m = k + 2p$ with $0 \leq p \in \mathbf{Z}$. Then for τ_ν and g as in (i) we have $(D_k^p\varphi)(\tau_1)/g(\tau_2) \in K_{\mathrm{ab}}$.*

PROOF. Clearly (i) implies (ii), as $D_k^p\varphi \in \mathscr{N}_m(\mathbf{Q}_{\mathrm{ab}})$; see Lemma 8.3 (i). We have $\tau_1 = a\tau_2 + b$ with $a, b \in \mathbf{Q}$. Put $h(z) = f(az + b)$. Then $h \in \mathscr{N}_m(\mathbf{Q}_{\mathrm{ab}})$ by Lemma 8.3 (iv), and $f(\tau_1)/g(\tau_2) = h(\tau_2)/g(\tau_2)$. This means that it is sufficient to prove (i) and also (ii) when $\tau_1 = \tau_2$. Thus we write hereafter τ for both τ_1 and τ_2. We first prove (ii) for $p = 1$ and $\varphi \in \mathscr{M}_k$. It is sufficient to prove the case where $\varphi(\tau) \neq 0$. Indeed, suppose $\varphi(\tau) = 0$. As noted above, we can find $\varphi_1 \in \mathscr{M}_k(\mathbf{Q})$ such that $\varphi_1(\tau) \neq 0$. Put then $\varphi_2 = \varphi + \varphi_1$. Then $\varphi_2(\tau) \neq 0$, and our assertion on $(D_k\varphi)(\tau)/g(\tau)$ follows from that on $(D_k\varphi_\nu)(\tau)/g(\tau)$ for $\nu = 1, 2$. Thus, assuming that $\varphi(\tau) \neq 0$, put $\psi = (\varphi\|_k\alpha)/\varphi$ as in §12.1. Then we obtain (12.2). Since $\varphi\|_k\alpha \in \mathscr{A}_k(\mathbf{Q}_{\mathrm{ab}})$ by Theorem 1.5, we see that $\psi \in \mathscr{A}_0(\mathbf{Q}_{\mathrm{ab}})$, and so $(2\pi i)^{-1}\psi' \in \mathscr{A}_2(\mathbf{Q}_{\mathrm{ab}})$. By (12.2) we have $(D_k\varphi/g)(\tau) = (\kappa^{-2} - 1)^{-1}\{(\varphi/g)\cdot(2\pi i)^{-1}\psi'/\psi\}(\tau)$, which belongs to K_{ab}, as $(\varphi/g)\cdot(2\pi i)^{-1}\psi'/\psi \in \mathscr{A}_0(\mathbf{Q}_{\mathrm{ab}})$. This proves (ii) for $p = 1$ and $\varphi \in \mathscr{M}_k$. Now $-24E_2 = \Delta^{-1}D_{12}\Delta$ by (8.18). Let r be an element of $\mathscr{M}_2(\mathbf{Q})$ such that $r(\tau) \neq 0$. Then $-24(E_2/r)(\tau) = \{D_{12}\Delta/(r\Delta)\}(\tau) \in K_{\mathrm{ab}}$ by the result just proved. Let f and g be as in (i). By Lemma 8.3 (ii), $f = \sum_{0\leq a\leq k/2} h_a E_2^a$ with $h_a \in \mathscr{M}_{k-2a}(\mathbf{Q}_{\mathrm{ab}})$. For $\tau \in K \cap H$, take r as above. The $(f/g)(\tau) = \sum_a (E_2/r)^a(\tau)(r^a h_a/g)(\tau) \in K_{\mathrm{ab}}$. This proves (i) and completes the proof.

As the above proof shows, $f(\tau)/g(\tau)$ for $f \in \mathscr{N}_m(\mathbf{Q}_{\mathrm{ab}})$ and $g \in \mathscr{A}_m(\mathbf{Q}_{\mathrm{ab}})$ can be reduced to the case where $f \in \mathscr{M}_m$ or $f = E_2$. Thus it is important to investigate the value $(E_2/r)(\tau)$ for $r \in \mathscr{M}_2(\mathbf{Q}_{\mathrm{ab}})$ such that $r(\tau) \neq 0$. More specifically we ask the following question:

Given τ, is there any canonical choice of r such that the value $(E_2/r)(\tau)$ can be determined explicitly?

The rest of this section is devoted to an answer to this question.

Lemma 12.3. (i) *Given $\alpha \in G_+$, put $\Gamma_\alpha = \alpha^{-1}\Gamma(1)\alpha \cap \Gamma(1)$ and define a function g_α on H by*

(12.3a) $$g_\alpha = E_2\|_2\alpha - E_2$$

with E_2 of (8.17). Then g_α is an Eisenstein series belonging to $\mathscr{M}_2(\mathbf{Q}_{ab}, \Gamma_\alpha)$, and

(12.3b) $$24g_\alpha = (2\pi i)^{-1}p^{-1}dp/dz, \quad p = (\Delta\|_{12}\alpha)^{-1}\Delta,$$

where Δ is as in (1.12). Moreover, put $h_m = 24g_\alpha$ when $\alpha = \mathrm{diag}[1, m]$ with $1 < m \in \mathbf{Z}$. Then $\Gamma_\alpha = \Gamma^0(m)$, h_m is an Eisenstein series belonging to $\mathscr{M}_2(\mathbf{Q}, \Gamma^0(m))$, and

(12.4a) $$h_m(z) = (24/m)E_2(z/m) - 24E_2(z),$$

(12.4b) $$h_{mn}(z) = h_n(z) + n^{-1}h_m(z/n) \qquad (1 < n \in \mathbf{Z}).$$

In particular, if m is a prime number, h_m is uniquely determined as an Eisenstein series belonging to $\mathscr{M}_2(\Gamma^0(m))$ whose Fourier expansion at ∞ has $(m-1)/m$ as its constant term. For $m = 2$, using the notation of (10.10a) and (10.21), we have

(12.5) $$h_2(z) = 6\varepsilon_1(z) = 6f(z;\, 1/2,\, 0).$$

PROOF. Clearly $g_\alpha\|_2\gamma = g_\alpha$ for every $\gamma \in \Gamma_\alpha$. By Lemma 8.3 (iv), $g_\alpha \in \mathscr{N}_2^1(\mathbf{Q}_{ab})$. We have $E_2 = (8\pi y)^{-1} + q$ with a holomorphic function q. Since $y \circ \alpha = |j_\alpha|^{-2}y$, we have $E_2\|_2\alpha = (8\pi y)^{-1}\overline{j}_\alpha j_\alpha^{-1} + q\|_2\alpha$. The term involving $(8\pi y)^{-1}$ can easily be seen to be $(8\pi y)^{-1}$ plus a holomorphic function, and so g_α is holomorphic. Also E_2 is orthogonal to $\mathscr{S}_2$, and so g_α is an Eisenstein series belonging to $\mathscr{M}_2(\mathbf{Q}_{ab}, \Gamma_\alpha)$. Next, $-24E_2\Delta = D_{12}\Delta$ by (8.18), and so, by (8.8),

(∗) $$-24(E_2\|_2\alpha)(\Delta\|_{12}\alpha) = (D_{12}\Delta)\|_{14}\alpha = D_{12}(\Delta\|_{12}\alpha).$$

With p as in (12.3b), we have $\Delta = (\Delta\|_{12}\alpha)p$. Therefore by (8.9), $D_{12}\Delta = D_{12}(\Delta\|_{12}\alpha)p + (\Delta\|_{12}\alpha)D_0p$. Thus, dividing this by $\Delta = (\Delta\|_{12}\alpha)p$, we obtain $-24E_2 = \Delta^{-1}D_{12}\Delta = (\Delta\|_{12}\alpha)^{-1}D_{12}(\Delta\|_{12}\alpha) + p^{-1}D_0p$. This combined with (∗) proves (12.3b). If $\alpha = \mathrm{diag}[1, m]$, then clearly $\Gamma_\alpha = \Gamma^0(m)$, and we obtain (12.4a), and so $h_m \in \mathscr{M}_2(\mathbf{Q})$. Formula (12.4b) follows immediately from (12.4a). If m is a prime number, then the number of inequivalent cusps of $\Gamma^0(m)$ is 2, and so the space of Eisenstein series contained in $\mathscr{M}_2(\Gamma^0(m))$ is one-dimensional, as can be seen from [S71, Theorems 2.23 and 2.24]. The expansion of h_m at ∞ can easily be described, and we obtain $(m-1)/m$ as its constant term. As for (12.5), we observe that ε_1 of (10.21) belongs to $\mathscr{M}_2(\Gamma^0(2))$. Therefore we obtain (12.5) by comparing the constant terms.

Theorem 12.4. (i) *Let $\tau \in K \cap H$ as in §12.1, and let α be an element of G_+ such that $\alpha(\tau) = \tau$ and $\alpha \notin \mathbf{Q}1_2$. Define g_α by (12.3a) and put $\kappa = j_\alpha(\tau)$. Then $\kappa^2 \neq 1$ and*

(12.6a) $$E_2(\tau) = (\kappa^{-2} - 1)^{-1}g_\alpha(\tau).$$

(ii) *Let m and a be integers such that $m > 1$ and $a^2 < 4m$; let τ be the root of the equation $\tau^2 - a\tau + m = 0$ belonging to H, that is, $\tau = (a + \sqrt{a^2 - 4m})/2$. Define h_m by (12.4a). Then*

$$24E_2(\tau) = (4m - a^2)^{-1}(a\tau - 2m)h_m(\tau). \tag{12.6b}$$

Moreover, $E_2(\tau) = 0$ for such a τ if and only if $4m - a^2 = 3$ or 4, that is, if and only if $\tau - (1 + \sqrt{-3})/2 \in \mathbf{Z}$ or $\tau - \sqrt{-1} \in \mathbf{Z}$. Consequently both $E_2(\tau)$ and $h_m(\tau)$ are nonzero in all the remaining cases.

PROOF. Evaluating (12.3a) at $z = \tau$, we obtain (12.6a). In §12.1 we have already seen that $\kappa^2 \neq 1$. With m and a as in (ii), put $\alpha = \begin{bmatrix} a & -m \\ 1 & 0 \end{bmatrix}$. Then $\alpha(\tau) = \tau$ and $E_2(\alpha(z)) = E_2(a - mz^{-1}) = E_2(-(z/m)^{-1}) = (z/m)^2 E_2(z/m)$, and so we obtain $(E_2\|_2\alpha)(z) = m^{-1}E_2(z/m)$. Thus $24g_\alpha$ coincides with h_m of (12.4a). Therefore (12.6a) shows that $24E_2(\tau) = (\kappa^{-2} - 1)^{-1}h_m(\tau)$ with $\kappa = j_\alpha(\tau) = m^{-1/2}\tau$. Since $1 - a\tau^{-1} + m\tau^{-2} = 0$, we have $\kappa^{-2} - 1 = m\tau^{-2} - 1 = a\tau^{-1} - 2$. Thus $(\kappa^{-2} - 1)^{-1} = \tau/(a - 2\tau) = \tau(a - 2\overline{\tau})/[(a - 2\tau)(a - 2\overline{\tau})] = (a\tau - 2m)/(4m - a^2)$, which gives (12.6b). To prove the second part of (ii), we first observe that $E_2(\zeta) = 0$ for $\zeta = i$ and $\zeta = \mathbf{e}(c/6)$ with $c = 1$ or 2. (Indeed, define q by (12.1) with ζ as τ and put $\varepsilon = q(\zeta)$. Then $\varepsilon(\zeta) = \zeta$ and $j_\varepsilon(\zeta) = \zeta$, and so $E_2(\zeta) = (E\|_2\varepsilon)(\zeta) = \zeta^{-2}E_2(\zeta)$.) Next, denote by S the last sum of (8.17) and put $t = e^{-2\pi y}$. Then $S = \sum_{n=1}^{\infty} n\mathbf{q}^n/(1 - \mathbf{q}^n)$ with $\mathbf{q} = \mathbf{e}(z)$, and so $|S| \leq \sum_{n=1}^{\infty} nt^n/(1-t) = t/(1-t)^3$. Suppose $y > 1.1$; then we easily find that $t < 0.001$ and $24^{-1} - (8\pi y)^{-1} > 0.005$, and so $E_2(z) \neq 0$ if $y > 1.1$. Thus, for τ as above, $E_2(\tau) = 0$ only if $4m - a^2 < 2.2^2$, that is, $0 < 4m - a^2 \leq 4$, as $4m - a^2$ is a positive integer. Now $4m = a^2 + n$ for $0 < n \leq 4$ is possible only for $n = 3$ or 4. If $n = 3$, then $\tau = (a + \sqrt{-3})/2$ and a is odd, and so, putting $\zeta = (1 + \sqrt{-3})/2$, we see that $\tau - \zeta \in \mathbf{Z}$, and so $E_2(\tau) = E_2(\zeta) = 0$. Similarly, if $n = 4$, then a is even and $\tau - i \in \mathbf{Z}$, and so $E_2(\tau) = E_2(i) = 0$. This completes the proof.

12.5. To state an easy example of (12.6b), take an arbitrary positive integer d. Then from (12.6b) with $a = 0$ and $m = d$ we obtain

$$24E_2(\sqrt{-d}) = (-1/2)h_d(\sqrt{-d}). \tag{12.7}$$

As another example, let $\tau = (1 + \sqrt{-d})/2$ with a positive integer d such that $d + 1 \in 4\mathbf{Z}$ and let $m = (d+1)/4$. Then (12.6b) gives

$$24E_2(\tau) = d^{-1}(\tau - 2m)h_m(\tau). \tag{12.8}$$

Let $K = \mathbf{Q}(\sqrt{-d_0})$ with a squarefree positive integer d_0, and let $\theta = \sqrt{-d_0}$ if $d_0 + 1 \notin 4\mathbf{Z}$, and let $\theta = (1 + \sqrt{-d_0})/2$ if $d_0 + 1 \in 4\mathbf{Z}$. Then the maximal order $\mathfrak{r}$ of K is given by $\mathfrak{r} = \mathbf{Z}\theta + \mathbf{Z}$. Thus (12.7) or (12.8) is applicable to $E_2(\theta)$.

It should be noted that $E_2(\theta)$ can be expressed in more than one way. For instance, assuming that $d_0 + 1 \notin 4\mathbf{Z}$, take $a = 2$ and $m = d_0 + 1$. Then τ of Theorem 12.4 equals $1 + \sqrt{-d_0}$. Then for $\theta = \sqrt{-d_0}$ we have

$$24E_2(\theta) = 24E_2(\theta + 1) = (2d_0)^{-1}(\theta - d_0)h_{d_0+1}(1 + \theta). \tag{12.8a}$$

This expression is different from (12.7).

Since there are nonprincipal ideals in K, it is natural to consider $E_2(\alpha/\beta)$ when $\mathbf{Z}\alpha + \mathbf{Z}\beta$ is a fractional ideal in K. To determine such a value, we first prove

Lemma 12.6. *Let σ be an element of $\mathfrak{r}$ such that $\mathfrak{r} = \mathbf{Z}\sigma + \mathbf{Z}$. Given an integral ideal $\mathfrak{x}$ in K, put $\{n \in \mathbf{Z} \mid n\sigma \in \mathfrak{x} + \mathbf{Z}\} = e\mathbf{Z}$ with $0 < e \in \mathbf{Z}$ and $\mathfrak{a} = e^{-1}\mathfrak{x}$. Then $\mathfrak{a}$ is an integral ideal and there exists an integer b such that*

$$\mathfrak{a} = \mathbf{Z}(\sigma + b) + \mathbf{Z}r, \quad r = N(\mathfrak{a}), \quad r \mid (\sigma + b)(\sigma^\rho + b). \tag{12.9}$$

Moreover, If $\mathfrak{c}$ is an integral ideal prime to r such that $\mathfrak{r} = \mathbf{Z} + \mathfrak{c}$, then b can be taken so that $\sigma + b \in \mathfrak{c}$.

PROOF. Take e as above for a given $\mathfrak{x}$ and put $\mathfrak{x} \cap \mathbf{Z} = a\mathbf{Z}$ with $0 < a \in \mathbf{Z}$; take $s \in \mathbf{Z}$ so that $e\sigma + s \in \mathfrak{x}$. If $t = f\sigma + g \in \mathfrak{x}$ with $f, g \in \mathbf{Z}$, then $e|f$. Putting $f = eh$, we see that $t - h(e\sigma + s) \in \mathfrak{x} \cap \mathbf{Z} = a\mathbf{Z}$. From this we can conclude that $\mathfrak{x} = \mathbf{Z}(e\sigma + s) + \mathbf{Z}a$. Put $\sigma^2 = k\sigma + \ell$ with $k, \ell \in \mathbf{Z}$. Since $\sigma\mathfrak{x} \subset \mathfrak{x}$ we have $\sigma(e\sigma + s) \in \mathfrak{x}$, and so $(s + ek)\sigma \in \mathfrak{x} + \mathbf{Z}$. Thus $e|s$. Similarly $\sigma a \in \mathfrak{x}$, and so $e|a$. Therefore $\mathfrak{a} = e^{-1}\mathfrak{x} = \mathbf{Z}(\sigma + b) + \mathbf{Z}r$ with $b = s/e$ and $r = a/e$. Since $\mathfrak{r} = \mathbf{Z}(\sigma+b)+\mathbf{Z}$, we have $N(\mathfrak{a}) = [\mathfrak{r} : \mathfrak{a}] = r$, and so r divides $(\sigma+b)(\sigma^\rho+b)$. This proves the first part. To prove the remaining part, take $\mathfrak{c}$ as $\mathfrak{x}$. Since $\mathfrak{r} = \mathfrak{c} + \mathbf{Z}$, we have $e = 1$, and so $\mathfrak{c} = \mathbf{Z}(\sigma + d) + \mathbf{Z}m$ with $d \in \mathbf{Z}$ and $m = N(\mathfrak{c})$. Since r is prime to m, we can put $d - b = mt + ru$ with $t, u \in \mathbf{Z}$. Put $b' = b + ru$. Then $\mathfrak{a} = \mathbf{Z}(\sigma + b') + \mathbf{Z}r$ and $\sigma + b' \in \mathfrak{c}$. This completes the proof.

12.7. Take θ of §12.5 as σ of Lemma 12.6; then every ideal class is represented by an ideal of the form $\mathfrak{a} = \mathbf{Z}(\theta + b) + \mathbf{Z}r$ with $b \in \mathbf{Z}$ and $r = N(\mathfrak{a})$. Thus our task is to find $E_2((\theta + b)/r)$. Taking m of (12.4a) to be r, we obtain $(24/r)E_2((\theta + b)/r) = 24E_2(\theta + b) + h_r(\theta + b) = 24E_2(\theta) + h_r(\theta + b)$. Thus

$$\frac{24E_2((\theta + b)/r)}{rh_m(\theta)} = \frac{24E_2(\theta)}{h_m(\theta)} + \frac{h_r(\theta + b)}{h_m(\theta)}, \tag{12.10}$$

where $m = (d_0 + 1)/4$ if $d_0 + 1 \in 4\mathbf{Z}$ and $m = d_0$ otherwise. The first quantity on the right-hand side is given by (12.7) or (12.8); the second quantity is the value of a modular function $h_r(z + b)/h_m(z)$ at $z = \theta$.

In Theorem A5.4 (v) of the Appendix we will investigate the behavior of a quantity of the type $f(\tau)/g(\tau)$ under $\mathrm{Gal}(K_{\mathrm{ab}}/K)$ for $\tau \in K \cap H$, $f \in \mathscr{N}_k(\mathbf{Q}_{\mathrm{ab}})$, and $g \in \mathscr{A}_k(\mathbf{Q}_{\mathrm{ab}})$. In particular, we will show in Theorem A5.6 that when these functions are $\mathbf{Q}$-rational and of level 1, then the value behaves like $J(\tau)$.

12.8. In Lemma 12.3 and Theorem 12.4 we considered modular forms with respect to $\Gamma^0(m)$. We can of course transform them to those with respect to $\Gamma_0(m)$ by putting

$$\ell_m(z) = h_m(-z^{-1})z^{-2}. \tag{12.11}$$

Then ℓ_m is an Eisenstein series belonging to $\mathscr{M}_2(\Gamma_0(m))$, and satisfies

$$\ell_m(z) = 24mE_2(mz) - 24E_2(z), \tag{12.12}$$

which follows immediately from (12.4a). However, this does not produce any formula better than (12.6b), which is why we formulated our results in terms of h_m. In any case, from (12.4a) and (12.12) we obtain $\ell_m(z) = -mh_m(mz)$, and so h_m satisfies

$$-mh_m(-mz^{-1})z^{-2} = h_m(z). \tag{12.13}$$

If τ is as in Theorem 12.4, then $m\tau^{-1} = \overline{\tau}$, and so by Lemma 1.6 (v) we obtain, excluding the cases in which $E_2(\tau) = 0$ as described in Theorem 12.4,

$$\overline{h_m(\tau)}/h_m(\tau) = -\tau^2/m = -\tau/\overline{\tau}. \tag{12.14}$$

13. The critical values of L-functions of an imaginary quadratic field

13.1. In Section 4 we discussed the values of the series

$$\sum_{0 \neq m \in q+N\mathbf{Z}} m^{-k}|m|^{-2s} \qquad (0 < k \in \mathbf{Z}).$$

If χ is a Dirichlet character modulo N such that $\chi(-1) = (-1)^k$, then $L(k, \chi)$ is a linear combination of such values. These values are π^k times numbers of $\mathbf{Q}_{\mathrm{ab}}$, and we stated various formulas for them in Theorems 4.7, 4.9, 4.12, 4.14, 4.16, 6.3, and 6.6. We also obtained recurrence formulas for these cyclotomic numbers in Theorem 4.15 and Section 11.

In this section we investigate analogues of these by considering the special values of the series introduced in §7.1. Thus we take the symbols K, $\mathfrak{r}$, and ρ as in §§7.1 and 12.1. Fixing a $\mathbf{Z}$-lattice $\mathfrak{b}$ in K and $\alpha \in K$, put $\alpha + \mathfrak{b} = \{\xi \in K \mid \xi - \alpha \in \mathfrak{b}\}$ and

$$L^\nu(s;\, \alpha,\, \mathfrak{b}) = \sum_{0 \neq \xi \in \alpha+\mathfrak{b}} \xi^{-\nu}|\xi|^{\nu-2s} \qquad (0 < \nu \in \mathbf{Z}), \tag{13.1}$$

where s is a complex variable as usual. In Theorem 7.3 we proved some analytic properties of this function. Since we assume $\nu > 0$ in this section, that theorem shows that our function of (13.1) can be continued to an entire function. Now we have

Theorem 13.2. *Let k be an integer such that $2-\nu \leq k \leq \nu$ and $\nu-k \in 2\mathbf{Z}$. Let $\sigma \in K \cap H$ and let g be an element of $\mathscr{M}_\nu(\mathbf{Q}_{\mathrm{ab}})$ such that $g(\sigma) \neq 0$. Then $L^\nu(k/2;\, \alpha,\, \mathfrak{b}) \in \pi^{(\nu+k)/2} g(\sigma) K_{\mathrm{ab}}$.*

PROOF. Put $\mathfrak{b} \cap \mathbf{Q} = c\mathbf{Z}$ with $c \in \mathbf{Q}^\times$. Then $\mathfrak{b}/c\mathbf{Z}$ is torsionfree. Replacing $(\alpha, \mathfrak{b})$ by $(c^{-1}\alpha, c^{-1}\mathfrak{b})$, we can reduce our problem to the case where $\mathfrak{b} = \mathbf{Z}\tau + \mathbf{Z}$ with $\tau \in K$. Changing τ for $-\tau$ if necessary, we may assume that $\tau \in H$. Take a positive integer N so that $N\alpha \in \mathfrak{b}$ and put $N\alpha = p\tau + q$ with $p, q \in \mathbf{Z}$. Then we easily see that

$$\text{(13.2)} \qquad \operatorname{Im}(\tau)^{s-\nu/2} N^{-2s} L^\nu(s;\, \alpha, \mathfrak{b}) = \mathfrak{E}_\nu^N(\tau,\, s - \nu/2;\, p,\, q).$$

Given k as in our theorem, put $n = (\nu - k)/2$. Then $n \in \mathbf{Z}$, $0 \le n \le \nu - 1$, and

$$\text{(13.2a)} \qquad L^\nu(k/2;\, \alpha, \mathfrak{b}) = \operatorname{Im}(\tau)^n N^k \mathfrak{E}_\nu^N(\tau,\, -n;\, p,\, q).$$

By Theorem 9.6 (iii), $\mathfrak{E}_\nu^N(z,\, -n;\, p,\, q)$ belongs to $\pi^{\nu-n}\mathscr{N}_\nu(\mathbf{Q}_{\mathrm{ab}})$. Therefore our assertion follows from Theorem 12.2 (i).

13.3. Our next problem is to find an explicit expression for the algebraic number $\pi^{-(\nu+k)/2} L^\nu(k/2;\, \alpha, \mathfrak{b})/g(\sigma)$ with a suitable choice of $g(\sigma)$ in the same sense as in Theorem 12.4. There is no clear-cut answer, but in many cases we can find a basic constant, and once we determine the values of $L^\nu(k/2;\, \alpha, \mathfrak{b})$ for a finite number of (k, ν), then the value for a more general (k, ν) can be reduced to those finitely many values by means of a recurrence formula, provided $k > 0$. (Here $(\alpha, \mathfrak{b})$ is fixed.)

To be explicit, suppose $\mathfrak{b} = \mathbf{Z}\tau + \mathbf{Z}$ with $\tau \in K \cap H$ and $N\alpha = p\tau + q \in \mathfrak{b}$ as in the proof of Theorem 13.2; suppose also $0 < k \le \nu$ and $\nu - k \in 2\mathbf{Z}$; put $n = (\nu - k)/2$, and

$$\text{(13.3)} \qquad \mathfrak{L}_k^n(\alpha, \mathfrak{b}) = (-1)^k (2\pi i)^{-k-n} \Gamma(k+n) L^\nu(k/2;\, \alpha, \mathfrak{b}) - \begin{cases} 2(2i)^n \operatorname{Im}(\tau)^n (D_2^n E_2)(\tau) & \text{if } k = 2 \text{ and } \alpha \notin \mathfrak{b}, \\ 0 & \text{otherwise.} \end{cases}$$

Then, assuming that $\alpha \notin \mathfrak{b}$, from (9.12), (10.10b, d), (10.13), and (13.2a) we obtain

$$\text{(13.4)} \qquad \mathfrak{L}_k^n(\alpha, \mathfrak{b}) = (2i)^n \operatorname{Im}(\tau)^n D_k^n F_k(\tau;\, p/N,\, q/N).$$

Take $g \in \mathscr{M}_1(\mathbf{Q})$ so that $g(\tau) \ne 0$. By Theorem 12.2 (ii), $\mathfrak{L}_k^n(\alpha, \mathfrak{b}) g(\tau)^{-k-2n}$ belongs to K_{ab}. Now, multiplying (10.17) by $(2i)^n \operatorname{Im}(\tau)^n g(\tau)^{-2n-r-5}$, we obtain a recurrence formula

$$\text{(13.5)} \qquad \frac{\mathfrak{L}_{r+5}^n(\alpha, \mathfrak{b})}{g(\tau)^{2n+r+5}} = 12 \sum_{i=0}^{r} \binom{r}{i} \sum_{j=0}^{n} \binom{n}{j} \frac{\mathfrak{L}_{i+3}^j(\alpha, \mathfrak{b})}{g(\tau)^{2j+i+3}} \cdot \frac{\mathfrak{L}_{r-i+2}^{n-j}(\alpha, \mathfrak{b})}{g(\tau)^{2n-2j+r-i+2}}$$

for $0 \le r \in \mathbf{Z}$ and $0 \le n \in \mathbf{Z}$, provided $\alpha \notin \mathfrak{b}$. By means of this formula the values of $\mathfrak{L}_k^n(\alpha, \mathfrak{b}) g(\tau)^{-k-2n}$ for $k > 4$ can be reduced inductively to those for $2 \le k \le 4$.

Next let us consider the case $\alpha \in \mathfrak{b}$. Clearly we can replace α by 0, and take $N = 1$. Thus

$$\mathfrak{L}_k^n(0,\,\mathfrak{b}) = (2i)^n\mathrm{Im}(\tau)^n(D_k^nE_k)(\tau) \qquad (0 < k \in 2\mathbf{Z}). \tag{13.6}$$

Applying D_{2n}^m to (10.8), using (8.9), and evaluating the result at $z = \tau$, we can state a recurrence formula for $\mathfrak{L}_k^n(0,\,\mathfrak{b})$ as follows:

$$\frac{(n-3)(2n+1)}{6(2n-2)!}\mathfrak{L}_{2n}^m(0,\,\mathfrak{b}) = \sum_{t=0}^{m}\binom{m}{t}\sum_{r,s}\frac{\mathfrak{L}_{2r}^t(0,\,\mathfrak{b})\mathfrak{L}_{2s}^{m-t}(0,\,\mathfrak{b})}{(2r-2)!(2s-2)!}, \tag{13.6a}$$

where $(r,\,s)$ is the same as in (10.8).

13.4. Returning to the original question about $L^\nu(k/2;\,\alpha,\,\mathfrak{b})$, we need to know $(D_2^nE_2)(\tau)$ in addition to $\mathfrak{L}_k^n(\alpha,\,\mathfrak{b})$. This question can be settled as follows. By (9.9a, b, c, d) and (10.8) we can find an expression for $D_2^nE_2$ as a polynomial of E_2, E_4, and E_6. Then the question can be reduced to their values at τ. The value $E_2(\tau)$ can be determined by Theorem 12.4. As for $E_4(\tau)$ and $E_6(\tau)$, they are essentially the parameters g_2 and g_3 associated to the curve $Y^2 = 4X^3 - g_2X - g_3$ isomorphic to $\mathbf{C}/\mathfrak{b}$, and obviously there is no general rule. We will compute $E_k(\tau)g(\tau)^{-k}$ for $k = 4,\,6$ in several examples we consider later.

In §13.3 we considered only the case $0 < k \le \nu$. Suppose now $2 - \nu \le k \le 0$. Taking k of (9.17b) to be ν and combining the result with (13.2a), we obtain

$$L^\nu(k/2;\,\alpha,\,\mathfrak{b}) = \mathrm{Im}(\tau)^nN^k\Gamma(\nu-n)^{-1}(-4\pi)^mD_h^m\mathfrak{E}_h^N(\tau,\,1-h;\,p,\,q), \tag{13.7}$$

where $n = (\nu-k)/2$, $m = (\nu+k-2)/2$, and $h = 2-k$. Thus the problem can be reduced to the function of (9.14), but there is no recurrence formula for such.

There is another way to deal with the case $2-\nu \le k \le 0$. Put $\ell = 2-k$ and take $r = \nu$ and $s = \ell/2$ in (7.8). Then

$$\pi^{-(k+\nu)/2}L^\nu(k/2;\,\alpha,\,\mathfrak{b}) = M\sum_\beta \mathbf{e}\big(2^{-1}\mathrm{Tr}_{K/\mathbf{Q}}(\beta\alpha)\big)\pi^{-(\ell+\nu)/2}L^\nu(\ell/2;\,\beta,\,\mathfrak{a}),$$

where $M = \Gamma\big((\ell+\nu)/2\big)\Gamma\big(1+(\nu-\ell)/2\big)^{-1}d(\mathfrak{b})^{-1/2}i^{-\nu}$ and $(\beta,\,\mathfrak{a})$ is as in (7.8). Since $2 \le \ell \le \nu$, our result of §13.3 is applicable to $L^\nu(\ell/2;\,\beta,\,\mathfrak{a})$.

13.5. We now fix an integral ideal $\mathfrak{c}$ and a nonzero integer ν, and consider a primitive or an imprimitive Hecke ideal character λ of K defined for the fractional ideals prime to $\mathfrak{c}$ such that

$$\lambda(\alpha\mathfrak{r}) = \alpha^{-\nu}|\alpha|^\nu \quad \text{if} \quad \alpha \in K^\times \quad \text{and} \quad \alpha - 1 \in \mathfrak{c}. \tag{13.8}$$

(This corresponds to a character λ^* of $K_\mathbf{A}^\times/K^\times$ such that $\lambda^*(x) = x^\nu|x|^{-\nu}$ for x in the archimedean completion of K.) Such a λ exists only if $\zeta^\nu = 1$ for every root of unity ζ such that $\zeta - 1 \in \mathfrak{c}$. Given such a λ, we easily see that there exists a character λ_0 of $(\mathfrak{r}/\mathfrak{c})^\times$ such that

$$\lambda(\alpha\mathfrak{r}) = \lambda_0(\alpha)\alpha^{-\nu}|\alpha|^\nu \text{ for every } \alpha \in \mathfrak{r} \text{ prime to } \mathfrak{c}. \tag{13.9}$$

Clearly $\lambda_0(\varepsilon) = \varepsilon^\nu$ for every $\varepsilon \in \mathfrak{r}^\times$. We define the L-function $L(s,\,\lambda)$ by

$$L(s,\,\lambda) = \textstyle\sum_\mathfrak{a}\lambda(\mathfrak{a})N(\mathfrak{a})^{-s},$$

where $\mathfrak{a}$ runs over all the integral ideals in K prime to $\mathfrak{c}$. If $\mathfrak{c}$ is the conductor of λ, then the functional equation for $L(s, \lambda)$ can be stated as follows:

$$\begin{aligned} R(s, \lambda) &:= |D_K N(\mathfrak{c})|^{s/2}(2\pi)^{1-(|\nu|/2)-s}\Gamma((s+(|\nu|/2))L(s, \lambda) \\ &= W(\lambda)R(1-s, \overline{\lambda}). \end{aligned}$$

Here D_K denotes the discriminant of K, and $W(\lambda)$ is an algebraic number of absolute value 1. Also, $R(s, \lambda)$ is an entire function. As to the explicit form of $W(\lambda)$ in terms of a generalized Gauss sum, the reader is referred to [S97, Theorem A6.3, formulas (A6.3.7) and (A6.3.8)].

Hereafter we assume that $\nu > 0$. (The case with negative ν can be reduced to this type by considering $\overline{\lambda}$.) For our problems it is practical and even advantageous not to assume that λ is primitive.

Let A be a complete set of representatives, consisting of integral ideals prime to $\mathfrak{c}$, for the ideal classes of K. Take a positive integer m such that $\mathfrak{c} \cap \mathbf{Z} = m\mathbf{Z}$. Since every integral ideal can be given in the form $\alpha\mathfrak{a}^{-1}$ with $\alpha \in \mathfrak{a} \cap K^\times$, we have

$$\begin{aligned} (13.10) \qquad wL(s, \lambda) &= \sum_{\mathfrak{a}\in A} \lambda(\mathfrak{a}^{-1})N(\mathfrak{a})^s \sum_{0\neq\alpha\in\mathfrak{a}} \lambda(\alpha\mathfrak{r})|\alpha|^{-2s} \\ &= \sum_{\mathfrak{a}\in A} \lambda(\mathfrak{a}^{-1})N(\mathfrak{a})^s \sum_{\xi\in\mathfrak{a}/m\mathfrak{a}} \lambda_0(\xi) \sum_{0\neq\alpha\in\xi+m\mathfrak{a}} \alpha^{-\nu}|\alpha|^{\nu-2s}, \end{aligned}$$

where w is the number of roots of unity in K and we put $\lambda_0(\xi) = 0$ for ξ not prime to $\mathfrak{c}$. The last sum over nonzero $\alpha \in \xi + m\mathfrak{a}$ can be written $L^\nu(s;\ \xi,\ m\mathfrak{a})$.

Now the basic result on the algebraicity of the special values of $L(s, \lambda)$ can be stated as follows:

Theorem 13.6. *Let k be an integer such that $2-\nu \le k \le \nu$ and $\nu-k \in 2\mathbf{Z}$. Let $\tau \in K \cap H$ and let h be an element of $\mathscr{M}_\nu(\overline{\mathbf{Q}})$ such that $h(\tau) \neq 0$. Then $L(k/2, \lambda)$ is an algebraic number times $\pi^{(\nu+k)/2}h(\tau)$.*

PROOF. From (13.10) we see that $L(k/2, \lambda)$ is a $\overline{\mathbf{Q}}$-linear combination of the quantitites of the type $L^\nu(k/2;\ \xi,\ m\mathfrak{a})$, and therefore our theorem follows immediately from Theorem 13.2.

13.7. As an easy example let us discuss the case in which $K = \mathbf{Q}(i)$ and $\mathfrak{c} = (1+i)^3\mathfrak{r}$. We have $E_k(i) = i^k E_k(i)$, and therefore

$$(13.11) \qquad E_k(i) = 0 \quad \text{if} \quad 0 \le k-2 \in 4\mathbf{Z}.$$

Since $(\mathfrak{r}/\mathfrak{c})^\times$ can be represented by $\mathfrak{r}^\times$, every integral ideal prime to $\mathfrak{c}$ is of the form $\alpha\mathfrak{r}$ with a unique α such that $\alpha - 1 \in \mathfrak{c}$. Thus there exists a unique character λ of conductor $\mathfrak{c}$ such that

$$(13.12) \qquad \lambda(\alpha\mathfrak{r}) = \alpha^{-1}|\alpha| \quad \text{if} \quad \alpha - 1 \in \mathfrak{c}.$$

We have therefore

$$L(s, \lambda^\nu) = L^\nu(s; 1, \mathfrak{c}) = \sum_{\beta\in\mathfrak{c}} (1+\beta)^{-\nu}|1+\beta|^{\nu-2s}. \tag{13.13}$$

Now $(1+i)\mathfrak{c} = 4\mathfrak{r}$, and so $(1+i)(1+\mathfrak{c}) = 1+i+4\mathfrak{r}$. Thus

$$(1+i)^{-\nu}2^{(\nu/2)-s}L(s, \lambda^\nu) = \mathfrak{E}_\nu^4(i, s-\nu/2; 1, 1). \tag{13.14}$$

Given $k \in \mathbf{Z}$ such that $2-\nu \le k \le \nu$ and $\nu - k \in 2\mathbf{Z}$, put $n = (\nu-k)/2$. Then

$$L(k/2, \lambda^\nu) = (1+i)^\nu 2^{-n}\mathfrak{E}_\nu^4(i, -n; 1, 1). \tag{13.14a}$$

Notice that

$$0 < k \le \nu \iff 0 \le n < \nu/2; \quad 2-\nu \le k \le 0 \iff \nu/2 \le n \le \nu - 1. \tag{13.14b}$$

Now $\mathfrak{E}_\nu^4(i, -n; 1, 1)$ can be given by (10.28). Therefore we obtain

$$L(k/2, \lambda^\nu) = (1-i)^\nu \pi^{(\nu+k)/2}\Gamma\big((\nu+k)/2\big)^{-1} \cdot \begin{cases} 2^{(\nu-3k)/2}D_k^n F_k^*(i;\, 1/4,\, 1/4) & \text{if } k > 0, \\ 2^{(\nu+k-4)/2}D_{2-k}^m F_{2-k}^{(4)}(i;\, 1/4,\, 1/4) & \text{if } k \le 0. \end{cases} \tag{13.15}$$

Here $F_k^* = F_k$ if $k \ne 2$ and $F_2^* = \mathscr{F}_2$; $m = (\nu+k-2)/2$.

13.8. Let us now compute $L(k/2, \lambda^\nu)$ numerically. The functions $F_k^*(z;\, 1/4,\, 1/4)$ and $F_k^{(4)}(z;\, 1/4,\, 1/4)$ belong to the group Γ_3 of §10.7, and so we denote the functions defined with $(a, b) = (1/4, 1/4)$ simply by F_k, $\mathscr{F}_2$, and $F_k^{(4)}$.

If $\omega_1/\omega_2 = z = i$, we have $\wp(iu) = -\wp(u)$, and so $e_2 = -e_1$ and $e_3 = 0$. Thus $\varepsilon_3(i) = 0$ and $\varepsilon_1(i)^2 = \varepsilon_2(i)^2 = 5E_4(i)$ by (10.22b). Therefore, from (10.26a, b, c, d) and (10.27a) we obtain

$$F_2(i) = \mathscr{F}_2(i) = 2F_1(i)^2, \quad F_3(i)/F_1(i)^3 = 8, \quad F_4(i)/F_1(i)^4 = 32,$$
$$E_4(i)/F_1(i)^4 = -4/5, \quad F_3^{(4)}(i)/F_1(i)^3 = 16. \tag{13.16}$$

Using these, we obtain the values $F_k(i)/F_1(i)^k$ for $k \ge 5$ from (10.15c). Similarly $(D^n E_k)(i)F_1(i)^{-k-2n}$ and $(D_k^n F_k)(i)F_1(i)^{-k-2n}$ can be obtained inductively from (9.9a, b, c, d), (10.26e, f, g), and (10.17). These are all rational numbers. The same can be said about $(D_k^n F_k^{(4)})(i)F_1(i)^{-k-2n}$ by virtue of Lemma 10.14. For instance,

$$(D_1F_1)(i) = 2F_1(i)^3, \quad (D_2\varepsilon_3)(i) = \tfrac{16}{3}F_1(i)^4, \quad (D_2F_2)(i) = \tfrac{16}{3}F_1(i)^4,$$
$$(D_3F_3)(i) = 16F_1(i)^5, \quad (D_5F_5)(i) = 2^7\cdot 7F_1(i)^7, \quad (D_2E_2)(i) = \tfrac{-2}{3}F_1(i)^4. \tag{13.17}$$

By (13.15), we can conclude that $L(k/2, \lambda^\nu)$ is $(1-i)^\nu\pi^{(\nu+k)/2}F_1(i)^\nu$ times a rational number that is computable. If $k > 0$, the recurrence formula (13.5) is applicable. For example,

$$L(1/2, \lambda) = 2^{-1}(1-i)\pi F_1(i), \quad L(5/2, \lambda^7) = (7/15)(1-i)^7\pi^6F_1(i)^7,$$
$$L(-1/2, \lambda^5) = -32(1-i)^5\pi^2F_1(i)^5. \tag{13.18}$$

To obtain the last quantity, we apply D_3 to (10.27a).

13.9. We can express such results in terms of the period of an elliptic integral. We have $\omega_1 = i\omega_2$, $g_2 = 4e_1^2$, and $g_3 = 0$, as $e_3 = 0$. Let ω denote the special choice of ω_2 for which $e_2 = 1$. In other words, we take ω so that $(2\pi i/\omega)^2\varepsilon_2(i) = 1$. Then $g_2 = 4$, and so, using the symbol of (10.4e), we have

$$C(i\omega,\, \omega)\ :\ Y^2 = 4X^3 - 4X \tag{13.19}$$

as the corresponding elliptic curve. Thus, from (10.20) we obtain

$$\omega = 2\int_1^\infty \frac{dx}{\sqrt{4x^3-4x}} = 2\int_0^1 \frac{dx}{\sqrt{1-x^4}}\,. \tag{13.20}$$

Now $4\omega^4 = g_2\omega^4 = (2\pi)^4\mathbf{g}_2(i) = 20(2\pi)^4E_4(i)$, and so

$$(2\pi/\omega)^4E_4(i) = 1/5. \tag{13.21}$$

Since $E_4(i)/F_1(i)^4 = -4/5$, we obtain $(2\pi/\omega)^4F_1(i)^4 = -1/4$, and so $F_1(i)$ is $(4\pi)^{-1}(1+i)\omega$ times a fourth root of unity. From the Fourier expansion of $F_1(z;\, 1/4,\, 1/4)$ we easily see that $\mathrm{Re}\big(F_1(i)\big)$ and $\mathrm{Im}\big(F_1(i)\big)$ are both positive. Thus we can conclude that

$$\omega = 2\pi(1-i)F_1(i;\, 1/4,\, 1/4) = 4L(1/2,\, \lambda). \tag{13.22}$$

Consequently, $L(k/2,\, \lambda^\nu)$ is $\pi^{(k-\nu)/2}\omega^\nu$ times a rational number that is computable.

It is also natural to compare ω with $F_1(i;\, 1/4,\, 0)$ and $F_1(i;\, 0,\, 1/4)$. Putting $z = i$ in (9.3c), we obtain $iF_1(i;\, 1/4,\, 0) = F_1(i;\, 0,\, 1/4)$. Put $\varphi(z) = F_1(z;\, 1/4,\, 0)$ and $x = (2\pi\varphi(i)/\omega)^2$. Since $(2\pi/\omega)^2\varepsilon_1(i) = -e_1 = e_2 = 1$, from (10.26d) with $\nu = 1$ and $(a,\, b) = (1/4,\, 0)$, we obtain $4x^2 - 6x = -1/4$. Solving this equation and comparing the result with the Fourier expansion of φ, we can conclude that

$$F_1(i;\, 1/4,\, 0) = -iF_1(i;\, 0,\, 1/4) = (4\pi)^{-1}(1+\sqrt{2}\,)\omega. \tag{13.23a}$$

$$\varepsilon_1(i)/F_1(i;\, 1/4,\, 0)^2 = \varepsilon_2(i)/F_1(i;\, 0,\, 1/4)^2 = 4(3-2\sqrt{2}\,). \tag{13.23b}$$

Call the curve of (13.19) simply C. In §14.19 we will show that $L(s,\, \lambda)$ is the zeta function of C over $\mathbf{Q}$, and also that there is a $\mathbf{Q}$-rational curve C' isomorphic to C over K but not over $\mathbf{Q}$. Thus the "periods" on C are diffferent from those on C. Still, $L(s,\, \lambda)$ is the zeta function of C' over $\mathbf{Q}$. Therefore the relationship between the periods and the critical values of $L(s,\, \lambda)$ is not so simple. This point will be investigated in Section 15B.

13.10. The conductor of λ^ν is $\mathfrak{r}$ if $\nu \in 4\mathbf{Z}$. In this case it is natural to consider $L(s,\, \mu^m)$ with a primitive character μ of conductor $\mathfrak{r}$ such that $\mu(\alpha\mathfrak{r}) = \alpha^{-4}|\alpha|^4$ for every $\alpha \in K^\times$. Indeed, for $0 < m \in \mathbf{Z}$ we have

$$4L(s,\, \mu^m) = \sum_{0\neq\alpha\in\mathfrak{r}} \alpha^{-4m}|\alpha|^{4m-2s} = \mathfrak{E}^1_{4m}(i,\, s-2m;\, 0,\, 0). \tag{13.24}$$

The quantity we should consider is $L(\kappa, \mu^m)$ with an integer κ such that $1 - 2m \le \kappa \le 2m$. Putting $n = 2m - \kappa$ and $n' = 2m + \kappa - 1$, from (10.28) and (9.22) we obtain

$$L(\kappa, \mu^m) = 2^{4m-1}\Gamma(2m+\kappa)^{-1}\pi^{2m+\kappa} \cdot \begin{cases} (D_{2\kappa}^n E_{2\kappa})(i) & (\kappa > 0), \\ (D_{2-2\kappa}^{n'} E_{2-2\kappa})(i) & (\kappa \le 0). \end{cases} \tag{13.25}$$

Since $E_2(i) = E_6(i) = 0$, the observation on $D_k^n E_k$ made in §9.2 shows that $(D_k^n E_k)(i)$ is $E_4(i)^{(k+2n)/4}$ times a rational number. Therefore, from (13.25) we see that $L(\kappa, \mu^m)$ is a rational number times $\pi^{\kappa-2m}\omega^{4m}$ in both cases $\kappa > 0$ and $\kappa \le 0$. For example, from (9.9a, b, c) and (10.9) we obtain $(D_4^2 E_4)(i) = (100/3)E_4(i)^2$, and so

$$L(-1, \mu^2) = 3^{-1}\pi^{-5}\omega^8. \tag{13.26}$$

13.11. In certain cases we obtain a simpler formula for $L(s, \lambda)$ than (13.10). Let λ, λ_0, $\mathfrak{c}$, and A be as in §13.5. We assume

$$\mathfrak{c} + \mathbf{Z} = \mathfrak{r}, \tag{13.27}$$

and put $N(\mathfrak{c}) = m$. Then $\mathbf{Z} \cap \mathfrak{c} = m\mathbf{Z}$ and $\mathfrak{r}/\mathfrak{c} \cong \mathbf{Z}/m\mathbf{Z}$. Take $\sigma_0 \in K \cap H$ so that $\mathfrak{r} = \mathbf{Z}\sigma_0 + \mathbf{Z}$. We apply Lemma 12.6 to the members of A with σ_0 as σ of that lemma. Take $\sigma_0 + b$ as in (12.9) and put $\sigma = \sigma_0 + b$. By the last assertion of Lemma 12.6, we may assume that each member $\mathfrak{a}$ of A is of the form

$$\mathfrak{a} = \mathbf{Z}\sigma + \mathbf{Z}r, \quad \sigma \in \mathfrak{c} \cap H, \quad r = N(\mathfrak{a}), \quad r \text{ is prime to } m. \tag{13.28}$$

Fix one $\mathfrak{a}$ of type (13.28). For $x\sigma + yr \in \mathfrak{a}$ with $x, y \in \mathbf{Z}$ we have $\lambda_0(x\sigma + yr) = \lambda_0(yr)$, and so

$$\sum_{0 \ne \alpha \in \mathfrak{a}} \lambda(\alpha\mathfrak{r})|\alpha|^{-2s} = \sum_{0 \ne (x,y) \in \mathbf{Z}^2} \lambda_0(yr)(x\sigma + yr)^{-\nu}|x\sigma + yr|^{\nu - 2s} \tag{13.29}$$
$$= \lambda_0(r)r^{-s-\nu/2}\mathrm{Im}(\sigma/m)^{\nu/2-s}\sum_{q=1}^{m}\lambda_0(q)\mathfrak{E}_\nu^m\big((mr)^{-1}\sigma,\, s - \nu/2;\, 0,\, q\big).$$

The last sum, up to some elementary factors, is the value of the function $\varphi_{\nu,\chi}^r$ of (9.10b) at $(mr)^{-1}\sigma$, where we take $N = m$ and χ to be the restriction of λ_0 to $\mathbf{Z}$; we assume that χ is primitive. We can replace $\varphi_{\nu,\chi}^r$ by $\varphi(z) = \varphi_{\nu,\chi}^r(z/m)$. Then the value in question becomes $\varphi(\sigma/r)$, and by (9.11g), $\varphi\|_\nu\gamma = \chi(a_\gamma)\varphi$ for every $\gamma \in \Gamma^0(N)$. In particular, if $\nu = 1$ and χ is real, then we have $\varphi = \varphi_{1,\chi}^\ell$ by (9.11h), and so

$$\chi\big(N(\mathfrak{a})\big)^{-1}N(\mathfrak{a})\Big\{\sum_{0 \ne \alpha \in \mathfrak{a}} \lambda(\alpha\mathfrak{r})|\alpha|^{-2s}\Big\}_{s=1/2} = \frac{4\pi}{\sqrt{m}}\varphi_{1,\chi}^\ell(\sigma/r). \tag{13.29a}$$

13.12. Let us consider an example of the setting of §13.11. Take $K = \mathbf{Q}(\sqrt{-7})$ and $\mathfrak{c} = \mathfrak{r}\tau^\rho$ with $\tau = (3+\sqrt{-7})/2$. Then $\mathfrak{r} = \mathbf{Z}\tau + \mathbf{Z}$, $N(\mathfrak{c}) = 4$, and there is a unique character λ of conductor $\mathfrak{c}$ such that $\lambda(\alpha\mathfrak{r}) = \lambda_0(\alpha)\alpha^{-1}|\alpha|$ for α prime to $\mathfrak{c}$ with λ_0 such that $\lambda_0(a) = \left(\dfrac{-4}{a}\right)$ for $a \in \mathbf{Z}$. Then λ^ν is primitive or imprimitive according as ν is odd or even.

Now (13.13) holds in the present case, and so instead of using (13.29), we can proceed as follows. Since $\tau(1+\mathfrak{c}) = \tau + 4\mathfrak{r}$, for the same reason as in (13.14) and (13.14a), we have

$$\tau^{-\nu}2^{\nu-2s}L(s,\,\lambda^\nu) = \operatorname{Im}(\tau)^{\nu/2-s}\mathfrak{E}_\nu^4(\tau,\,s-\nu/2;\,1,\,0), \tag{13.30}$$

$$L(k/2,\,\lambda^\nu) = \tau^\nu 2^{-3n}7^{n/2}\mathfrak{E}_\nu^4(\tau,\,-n;\,1,\,0). \tag{13.30a}$$

Here $2-\nu \le k \le \nu$ and $n = (\nu-k)/2 \in \mathbf{Z}$. By (10.28) we obtain

$$L(k/2,\,\lambda^\nu) = (-i\tau)^\nu 7^{(\nu-k)/4}\Gamma\big((\nu+k)/2\big)^{-1}\pi^{(\nu+k)/2} \cdot \begin{cases} 2^{-(\nu+k)/2}D_k^nF_k^*(\tau;\,1/4,\,0) & (k>0),\\ 2^{(3k-\nu-4)/2}D_{2-k}^mF_{2-k}^{(4)}(\tau;\,1/4,\,0) & (k\le 0),\end{cases} \tag{13.30b}$$

where the symbols are as in (10.28). Thus we can naturally take $F_1(\tau;\,1/4,\,0)$ as a basic constant. However, (13.29) is also useful. In order to use it, we take $\sigma = -\tau^\rho = (-3+\sqrt{-7})/2$. Then for the present λ we have, by (13.29a),

$$L(1/2,\lambda) = \pi F_1(\sigma;\,1/4,\,0). \tag{13.30c}$$

This combined with (13.30b) shows that

$$F_1(\sigma;\,1/4,\,0) = (-i\tau/2)F_1(\tau;\,1/4,\,0). \tag{13.30d}$$

Therefore we can also take $F_1(\sigma;\,1/4,\,0)$ as a basic constant. In the next subsection we determine various quantities in terms of these constants.

13.13. Let $w = 20E_4(z)\varepsilon_\nu(z)^{-2}$. By (10.33) we have $J(z)(w-3)(w-12)^2 = w^3$. Therefore, once the value of $J(z)$ is given, we can find the values of the three functions $20E_4(z)\varepsilon_\nu(z)^{-2}$ by solving a cubic equation.

Let $\theta = (1+\sqrt{-7})/2$ and $\tau = (3+\sqrt{-7})/2$. Since $\tau = \theta + 1$, from (10.23a) we obtain $\varepsilon_2(\tau) = \varepsilon_2(\theta)$, $\varepsilon_1(\tau) = \varepsilon_3(\theta)$, and $\varepsilon_3(\tau) = \varepsilon_1(\theta)$; of course $E_{2m}(\tau) = E_{2m}(\theta)$. From (12.5) and (12.6b) we obtain, taking $a=1$ and $m=2$,

$$E_2(\tau)/\varepsilon_3(\tau) = E_2(\theta)/\varepsilon_1(\theta) = (\sqrt{-7}-7)/56. \tag{13.31}$$

Now $J(\theta) = J(\tau) = -5^3/4^3$, which follows from [Web, p. 179, (4) and p. 460, (18)], and so $20E_4(\tau)\varepsilon_\nu(\tau)^{-2}$ satisfies an equation

$$5^3(w-3)(w^2-24w+144) + 4^3w^3 = 0,$$

which can be transformed to $7w^3 - 5^3(w-4)^2 = 0$. It is not so painful to find that the roots are $20/7$ and $5(3 \pm \sqrt{-7})/2$. From the Fourier expansion of each function, we see that $E_{2n}(\tau)$ and $\varepsilon_2(\tau)$ are real. From (13.31) we see that $\varepsilon_3(\tau)^{-2}$ is a real number times $3 - \sqrt{-7}$, and therefore we obtain the first three of the following equalities:

(13.32a) $$E_4(\tau)/\varepsilon_1(\tau)^2 = (3+\sqrt{-7})/8,$$

(13.32b) $$E_4(\tau)/\varepsilon_2(\tau)^2 = 1/7,$$

(13.32c) $$E_4(\tau)/\varepsilon_3(\tau)^2 = (3-\sqrt{-7})/8,$$

(13.32d) $$\varepsilon_1(\tau) = \overline{\varepsilon_3(\tau)},$$

(13.32e) $$\varepsilon_1(\tau)/F_1(\tau;\, 1/4,\, 0)^2 = 1-\sqrt{-7},$$

(13.32f) $$\varepsilon_2(\tau)/F_1(\tau;\, 0,\, 1/4)^2 = 4\sqrt{7}(8-3\sqrt{7}),$$

(13.32g) $$\varepsilon_3(\tau)/F_1(\tau;\, 1/4,\, 1/4)^2 = 1+\sqrt{-7},$$

(13.32h) $$E_6(\tau)/\varepsilon_1(\tau)^3 = -3(7+5\sqrt{-7})/14,$$

(13.32i) $$E_6(\tau)/\varepsilon_2(\tau)^3 = 24/49,$$

(13.32j) $$E_6(\tau)/\varepsilon_3(\tau)^3 = -3(7-5\sqrt{-7})/14,$$

(13.32k) $$\varepsilon_2(\tau)/\varepsilon_1(\tau) = (-7-\sqrt{-7})/4,$$

(13.32l) $$\varepsilon_2(\tau)/\varepsilon_3(\tau) = (-7+\sqrt{-7})/4,$$

(13.32m) $$E_2(\tau)/\varepsilon_2(\tau) = 1/14.$$

To prove (13.32d), we note that $\varepsilon_3 \in \mathscr{M}_2(\mathbf{R})$. Since $\tau + \overline{\tau} = 3$, by (10.23a) and Lemma 1.6 (v) we have $\varepsilon_1(\tau) = \varepsilon_1(3-\overline{\tau}) = \varepsilon_3(-\overline{\tau}) = \overline{\varepsilon_3(\tau)}$ as expected. Next, put $x = F_1(\tau;\, 1/4,\, 0)^2/\varepsilon_1(\tau)$. From (10.26d) and (13.32a) we obtain $4x^2 - 6x - 3/4 + (5/8)(3+\sqrt{-7}) = 0$. The solutions are $(11-\sqrt{-7})/8$ and $(1+\sqrt{-7})/8$. Using the Fourier expansions of F_1 and ε_1, we easily find a good numerical approximation to x. Thus we obtain (13.32e) and also (13.32f, g) in a similar way. Equalities (13.32h, i, j) follow from (10.26e) and (13.32a, b, c); (13.32k, l) follow from (13.32a, b, c) and (13.32h, i, j); then (13.32m) follows from (13.31) and (13.32l).

By Lemma 10.14 and (10.26a), $F_k(z;\, a_\nu,\, b_\nu)$ and $F_k^{(4)}(z;\, a_\nu,\, b_\nu)$ are polynomials in $F_1(z;\, a_\nu,\, b_\nu)$ and $\varepsilon_\nu(z)$ with rational coefficients. Therefore in view of (10.26f, g) and (9.9a, b, c) we see that $D_k^n F_k^*(z;\, a_\nu,\, b_\nu)$ and $D_k^n F_k^{(4)}(z;\, a_\nu,\, b_\nu)$ are polynomials in $F_1(z;\, a_\nu,\, b_\nu)$, $\varepsilon_\nu(z)$, and $E_2(z)$ with rational coefficients. Consequently these quantities are powers of $F_1(\tau;\, a_\nu,\, b_\nu)$ times elements of $K = \mathbf{Q}(\sqrt{-7})$ if $\nu \neq 2$; they are powers of $F_1(\tau;\, 0,\, 1/4)$ times elements of $\mathbf{Q}(\sqrt{7})$ if $\nu = 2$. Thus from (13.30b, d) we see that $L(k/2,\, \lambda^\nu)$ is $\pi^{(k+\nu)/2} F_1(\sigma;\, 1/4,\, 0)^\nu$ times a number of K. For instance,

(13.33) $$L(-1/2,\, \lambda^3) = 14(3\sqrt{-7}-1)\pi F_1(\sigma;\, 1/4,\, 0)^3.$$

13.14. If $\nu = 2m \in 2\mathbf{Z}$ in the setting of §13.12, it is natural to consider μ^m instead of λ^{2m}, where μ is the primitive character of conductor $\mathfrak{r}$ such that $\mu(\alpha\mathfrak{r}) = \alpha^{-2}|\alpha|^2$ for every $\alpha \in K^\times$. We then ask the nature of $L(\kappa, \mu^m)$ for an integer κ such that $1 - m \le \kappa \le m$. We can easily state analogues of (13.24) and (13.25), whose explicit forms we leave to the reader. In this case it is natural to take $\varepsilon_2(\tau)$ as a basic constant. We eventually find that $L(\kappa, \mu^m)$ is $7^{(m-\kappa)/2}\pi^{m+\kappa}\varepsilon_2(\tau)^m$ times a rational number.

Let us now consider the Weierstrass equation for $\omega_1/\omega_2 = \tau = (3+\sqrt{-7})/2$. For each choice of ω_2 we obtain an equation that determines an elliptic curve isomorphic to $\mathbf{C}/(\mathbf{Z}\tau + \mathbf{Z})$. One natural choice is $\omega_2 = 2\pi i|\varepsilon_2(\tau)|^{1/2}$. (Notice that $0 > \varepsilon_2(\tau) \in \mathbf{R}$.) Then from (13.32b, i) we see that the curve has the form

$$(13.34) \qquad C(\tau\omega_2, \omega_2) : Y^2 = 4X^3 - (20/7)X + 8/7, \quad \omega_2 = 2\pi i|\varepsilon_2(\tau)|^{1/2}.$$

It is also natural to take $\sigma = -\tau^\rho$ instead of τ, as (13.30c) suggests. Put $\varphi(z) = F_1(z;\, 1/4, 0)$ for simplicity. Then from (13.30d) and (13.32e, k) we obtain $\varepsilon_2(\tau) = \sqrt{-7}\,\sigma\varphi(\sigma)^2$. Therefore the curve $C(\sigma\omega, \omega)$ with $\omega = 2\pi i\varphi(\sigma)$ has the form

$$(13.34a) \qquad C(\sigma\omega, \omega) : Y^2 = 4X^3 - (20/7)\gamma^2 X + (8/7)\gamma^3, \quad \gamma = \sqrt{-7}\,\sigma.$$

Notice that $\sigma = (1+\sqrt{-7})^2/4$.

13.15. We now take $K = \mathbf{Q}(\zeta)$, $\zeta = \mathbf{e}(1/3)$, and $\mathfrak{c} = 3\mathfrak{r}$. There is a unique ideal character λ of conductor $\mathfrak{c}$ satisfying (13.12), as $(\mathfrak{r}/\mathfrak{c})^\times$ can be represented by $\mathfrak{r}^\times$. Then (13.13) is also valid in the present case. Therefore $\mathrm{Im}(\zeta)^{s-\nu/2}L(s, \lambda^\nu) = \mathfrak{E}^3_\nu(\zeta, s - \nu/2;\, 0, 1)$. Thus, from (10.28) we obtain

$$(13.35) \qquad L(k/2, \lambda^\nu) = (-i)^\nu \Gamma\big((\nu+k)/2\big)^{-1}(2\pi)^{(\nu+k)/2} \cdot \begin{cases} 3^{(n-2k)/2}D_k^n F_k^*(\zeta;\, 0, 1/3) & (k > 0), \\ 3^{(n-2)/2}D_{2-k}^m F_{2-k}^{(3)}(\zeta;\, 0, 1/3) & (k \le 0), \end{cases}$$

Here $2 - \nu \le k \le \nu$, $n = (\nu - k)/2 \in \mathbf{Z}$, $m = (\nu + k - 2)/2$; $F_k^* = F_k$ if $k \ne 2$ and $F_2^* = \mathscr{F}_2$.

We can replace $(0, 1/3)$ by other vectors. For example, let $\sigma = -b - \zeta^2$ with $b \in \mathbf{Z}$ and $\alpha = \begin{bmatrix} 0 & -1 \\ 1 & b \end{bmatrix}$. Then $\sigma \in H$, $\mathfrak{r} = \mathbf{Z}\sigma + \mathbf{Z}$, and $\zeta = \alpha(\sigma)$. From (9.3) we obtain $(-\zeta)^k F_k(\zeta;\, 0, 1/3) = F_k(\sigma;\, 1/3, b/3)$.

Now let τ be an element of $K \cap H$ such that $\mathfrak{r} = \mathbf{Z}\tau + \mathbf{Z}$. Then $E_k(\tau) = E_k(\zeta) = 0$ for $k \notin 6\mathbf{Z}$, as $E_k(\zeta) = \zeta^k E_k(\zeta)$. Let F_k be as in §10.8. Then from (10.29a, b, d) we obtain

$$(13.36) \quad F_2(\tau) = \mathscr{F}_2(\tau) = 3F_1(\tau)^2, \quad F_3(\tau) = 3^2F_1(\tau)^3, \quad E_6(\tau) = (3^4/7)F_1(\tau)^6,$$

assuming that $F_1(\tau) \neq 0$. We can easily verify that $F_1(\zeta;\,1/3,\,0)$, $F_1(\zeta;\,0,\,1/3)$, and $F_1(\zeta,\,1/3,\,1/3)$ are all nonzero, but $F_1(\zeta;\,1/3,\,2/3) = 0$. By Lemma 10.14, (10.29a–f), and (13.35) we can conclude that $L(k/2,\,\lambda^\nu)$ is $3^{(\nu-k)/4}\pi^{(\nu+k)/2}i^\nu \cdot F_1(\zeta;\,0,\,1/3)^\nu$ times a rational number.

Next, we take $\omega_1 = \zeta\omega_2$ in the setting of §10.6. Clearly $g_2 = 0$, $\wp(\zeta u) = \zeta\wp(u)$, $\omega_1 + \omega_2 = (\zeta+1)\omega_2 = -\zeta^2\omega_2$, and so $e_1 = \zeta e_2$ and $e_3 = \zeta^2 e_2$, and $g_3 = 4e_1e_2e_3 = 4e_2^3$. Let ω be the value of ω_2 for which e_2 is real and $4e_2^3 = 3^3$. Then $(2\pi/\omega)^6 E_6(\zeta) = -3^4/7$ by (10.4b), and the corresponding elliptic curve is

$$(13.37) \qquad C(\zeta\omega,\,\omega)\ :\ Y^2 = 4X^3 - 3^3.$$

By (10.20) we have

$$(13.38) \qquad \omega = 2^{1/3}3^{-1/2}\int_1^\infty \frac{dx}{\sqrt{x^3-1}}.$$

Now $\left(2\pi F_1(\zeta)/\omega\right)^6 = -1$ by the last equality of (13.36). The function $F_1(z;\,0,\,1/3)$ is $i\sqrt{3}$ times the function of (9.11b) with $N = 3$ and $k = 1$. From its Fourier expansion we see that $F_1(\zeta;\,0,\,1/3)$ is i times a positive real number, and so

$$(13.39) \qquad \omega = -2\pi i F_1(\zeta;\,0,\,1/3) = 3L(1/2,\,\lambda).$$

Employing this, we see that $L(k/2,\,\lambda^\nu)$ is $3^{(\nu-k)/4}\pi^{(k-\nu)/2}\omega^\nu$ times a rational number that is computable. For instance, by (9.9a) and (10.29a) we have $D_2\mathscr{F}_2 = (5/3)E_4 - 4E_2^2 + 6F_1D_1F_1$, and so from (10.29e) and (13.36) we obtain $(D_2\mathscr{F}_2)(\zeta) = 9F_1(\zeta)^4$. Thus

$$(13.40) \qquad L(1,\lambda^4) = 3^{1/2}(4\pi)^{-1}\omega^4 = 3^{1/2}\cdot 4\pi^3 F_1(\zeta;\,0,\,1/3)^4.$$

13.16. Next we take $K = \mathbf{Q}(\sqrt{-2}\,)$ and $\mathfrak{c} = 2\sqrt{-2}\,\mathfrak{r}$. To make our formulas short, let us put $\sigma = \sqrt{-2}$. Then $(\mathfrak{r}/\mathfrak{c})^\times$ is a cyclic group of order 4 generated by $1+\sigma$ modulo $\mathfrak{c}$. Thus we have a character λ satisfying

$$(13.41) \qquad \lambda(\alpha\mathfrak{r}) = \lambda_0(\alpha)\alpha^{-1}|\alpha|$$

for α prime to $\mathfrak{c}$ with a character λ_0 of $(\mathfrak{r}/\mathfrak{c})^\times$ such that $\lambda_0(1+\sigma) = i$. Since $\sigma(1+\mathfrak{c}) = \sigma + 4\mathfrak{r}$, by the same technique as in (13.14) we easily see, for $0 < \nu \in \mathbf{Z}$, that

$$i^{-\nu}2^{-s}L(s,\,\lambda^\nu) = L^\nu(s;\,\sigma,\,4\mathfrak{r}) + i\cdot L^\nu(s;\,\sigma-2,\,4\mathfrak{r}).$$

Therefore, given an integer k such that $2-\nu \le k \le \nu$ and $\nu - k = 2n \in 2\mathbf{Z}$, we have

$$(13.42) \qquad L(k/2,\,\lambda^\nu) = i^\nu 2^{(k+\nu)/4}\bigl\{\mathfrak{E}_\nu^4(\sigma,-n;\,1,\,0) + i\cdot\mathfrak{E}_\nu^4(\sigma,-n;\,1,\,2)\bigr\}.$$

Thus we need $F_k(z;\,1/4,\,1/2)$ in addition to $F_k(z;\,1/4,\,0)$; also, (10.28) is applicable. To simplify our notation, we put

$$(13.43) \qquad \varphi_k(z) = F_k(z;\,1/4,\,1/2).$$

Since ε_1 and E_k are invariant under $z \mapsto z+2$, we see that formulas (10.26a–g) are valid with φ_k, ε_1, and $\mathscr{F}_2(z;\, 1/4,\, 1/2)$ in place of F_k, ε_ν, and $\mathscr{F}_2$. In particular, from (10.26d) we obtain $4(\varphi_1^4 - F_1^4) = 6(\varphi_1^2 - F_1^2)\varepsilon_1$, and so

$$2\varphi_1(z)^2 + 2F_1(z;\, 1/4,\, 0)^2 = 3\varepsilon_1(z). \tag{13.44}$$

We have also $2\varphi_2 = 2\varphi_1^2 - \varepsilon_1$ and $\varphi_3 = 8\varphi_1^3 - 6\varphi_1\varepsilon_1$. We note that $J(\sigma) = 5^3/3^3$, as can be seen from [Web, p. 460, (17)]. Then by the same technique as in §13.13 we obtain the following numerical results:

$$E_2(\sigma)/\varepsilon_1(\sigma) = -1/8, \tag{13.45a}$$

$$E_4(\sigma)/\varepsilon_1(\sigma)^2 = 3/8, \tag{13.45b}$$

$$E_4(\sigma)/\varepsilon_2(\sigma)^2 = 3(11 - 6\sqrt{2})/49, \tag{13.45c}$$

$$E_4(\sigma)/\varepsilon_3(\sigma)^2 = 3(11 + 6\sqrt{2})/49, \tag{13.45d}$$

$$E_6(\sigma)/\varepsilon_1(\sigma)^3 = 6/7, \tag{13.45e}$$

$$\varepsilon_1(\sigma)/F_1(\sigma;\, 1/4,\, 0)^2 = 4(2 - \sqrt{2})/3, \tag{13.45f}$$

$$\varphi_1(\sigma)/F_1(\sigma;\, 1/4,\, 0) = \sqrt{2} - 1, \tag{13.45g}$$

$$F_3(\sigma;\, 1/4,\, 0)/F_1(\sigma;\, 1/4,\, 0)^3 = 8(\sqrt{2} - 1), \tag{13.45h}$$

$$\varphi_3(\sigma)/F_1(\sigma;\, 1/4,\, 0)^3 = 8(2\sqrt{2} - 3). \tag{13.45i}$$

Therefore, from (13.42) we obtain

$$L(1/2,\, \lambda)/F_1(\sigma;\, 1/4,\, 0) = \pi(\sigma + 1 - i)\sqrt{2}. \tag{13.46}$$

We now consider the elliptic curve of (10.3b) with $(\omega_1,\, \omega_2) = (\sigma\omega,\, \omega)$, $\omega = 2\pi i \varepsilon_1(\sigma)^{1/2}$. (Notice that $\varepsilon_1(\sigma) > 0$.) Then $g_2(\omega_1,\, \omega_2) = 20E_4(\sigma)/\varepsilon_1(\sigma)^2 = 15/2$ and $g_3(\omega_1,\, \omega_2) = (7/3)E_6(\sigma)/\varepsilon_1(\sigma)^3 = 2$. Thus $\mathbf{C}/\mathfrak{r}$ is isomorphic to

$$C(\sigma\omega,\, \omega) \,:\, Y^2 = 4X^3 - (15/2)X - 2, \quad \omega = 2\pi i \varepsilon_1(\sigma)^{1/2}. \tag{13.47}$$

13.17. In the above discussion we chose a quantity of the form $F_1(\tau;\, a,\, b)$ as a basic constant. Instead, we can take $\eta(\tau)^2 = \varDelta(\tau)^{1/12}$, but this may not be the best choice. Still it is natural to ask how it is related to $F_1(\tau;\, a,\, b)$. This can be answered as follows. Once $E_4(\tau)F_1(\tau;\, a,\, b)^{-4}$ and $E_6(\tau)F_1(\tau;\, a,\, b)^{-6}$ are determined, we obtain $\varDelta(\tau)F_1(\tau;\, a,\, b)^{-12}$ from (10.4c), and that gives the value of $\eta(\tau)^2/F_1(\tau;\, a,\, b)$ up to a twelfth root of unity. Since a numerical approximation of $\eta(\tau)^2$ can easily be done, we eventually obtain a numerical value of $\eta(\tau)^2/F_1(\tau;\, a,\, b)$. For example, we have

$$\eta(i)^2 = \sqrt{2}(1 - i)F_1(i;\, 1/4,\, 1/4), \tag{13.48a}$$

$$\eta(\zeta)^2 = \mathbf{e}(-7/24)F_1(\zeta;\, 0,\, 1/3), \qquad \zeta = \mathbf{e}(1/3). \tag{13.48b}$$

14. The zeta function of a member of a one-parameter family of elliptic curves

14.1. In this subsection we let F, $\mathfrak{g}$, $\mathbf{a}$, and $\mathbf{h}$ denote an algebraic number field of finite degree, the ring of algebraic integers in F, the set of all archimedean primes of F, and the set of all nonarchimedean primes of F, respectively. For every $v \in \mathbf{a} \cup \mathbf{h}$ we denote by F_v the v-completion of F. In particular, for $v \in \mathbf{h}$ and a fractional ideal $\mathfrak{x}$ we denote by $\mathfrak{x}_v$ the v-closure of $\mathfrak{x}$ in F_v, which coincides with the $\mathfrak{g}_v$-linear span of $\mathfrak{x}$ in F_v. If $F = \mathbf{Q}$ in particular, then $\mathfrak{g} = \mathbf{Z}$, and we use a prime number p instead of v. Thus the subscript p means the p-completion, so that $\mathbf{Q}_p$ and $\mathbf{Z}_p$ mean the field of p-adic numbers and the ring of p-adic integers. Then we put $F_p = F \otimes_{\mathbf{Q}} \mathbf{Q}_p$. This can be identified with $\prod_{v|p} F_v$. For any $\mathbf{Z}$-lattice $\mathfrak{a}$ in F we denote by $\mathfrak{a}_p$ the $\mathbf{Z}_p$-linear span of $\mathfrak{a}$ in F_p. If $\mathfrak{a}$ is a fractional ideal, then $\mathfrak{a}_p$ can be identified with $\prod_{v|p} \mathfrak{a}_v$.

We denote by $F_{\mathbf{A}}$, $F_{\mathbf{A}}^{\times}$, $F_{\mathbf{a}}$, and $F_{\mathbf{h}}$ the adele ring of F, the idele group of F, the archimedean factor of $F_{\mathbf{A}}$, and the nonarchimedean factor of $F_{\mathbf{A}}$. For $x \in F_{\mathbf{A}}$ and $v \in \mathbf{a} \cup \mathbf{h}$, we denote by x_v the v-component of x; also, we denote by x_p, $x_{\mathbf{a}}$, and $x_{\mathbf{h}}$ the projections of x to F_p, $F_{\mathbf{a}}$, and $F_{\mathbf{h}}$. Given $y \in F_{\mathbf{A}}^{\times}$ and a $\mathbf{Z}$-lattice $\mathfrak{a}$ in F, we denote by $y\mathfrak{a}$ the $\mathbf{Z}$-lattice in F such that $(y\mathfrak{a})_p = y_p\mathfrak{a}_p$ for every prime number p. If $\mathfrak{a}$ is a fractional ideal, then $y\mathfrak{a}$ is a fractional ideal. We also put $|y|_{\mathbf{A}} = \prod_v |y_v|_v$, where the product is taken over all $v \in \mathbf{a} \cup \mathbf{h}$, and $| \ |_v$ is the normalized valuation at v. Notice that $N(y\mathfrak{g}) = |y|_{\mathbf{A}}^{-1}$ if $y \in F_{\mathbf{h}}^{\times}$. For a matrix X with entries in $F_{\mathbf{A}}$ and a fractional ideal $\mathfrak{x}$ in F we write $X \prec \mathfrak{x}$ if all the entries of X_v are contained in $\mathfrak{x}_v$ for every $v \in \mathbf{h}$.

We let F_{ab} denote the maximal abelian extension of F. (When we fix an embedding of F in $\mathbf{C}$ as we will do later, we take F_{ab} to be the maximal abelian extension of F in $\mathbf{C}$.) By class field theory there is a canonical surjective homomorphism of $F_{\mathbf{A}}^{\times}$ onto $\mathrm{Gal}(F_{\mathrm{ab}}/F)$. For $y \in F_{\mathbf{A}}^{\times}$ and $a \in F_{\mathrm{ab}}$ we denote by $\{y\}$ the image of y under that homomorphism, and by $a^{\{y\}}$ the image of a under $\{y\}$.

By a **Hecke character** of $F_{\mathbf{A}}^{\times}$ we understand a continuous homomorphism ψ of $F_{\mathbf{A}}^{\times}$ into $\mathbf{C}^{\times}$ such that $\psi(F^{\times}) = 1$ and $|\psi(x)| = 1$ for every $x \in F_{\mathbf{A}}^{\times}$. For such a ψ we denote by ψ_v, $\psi_{\mathbf{a}}$, and $\psi_{\mathbf{h}}$ its restrictions to $F_v^{\times}$, $F_{\mathbf{a}}^{\times}$, and $F_{\mathbf{h}}^{\times}$, respectively. Given ψ, there exists a unique integral ideal $\mathfrak{f}$ with the following property: $\psi_v(x) = 1$ if $v \in \mathbf{h}$, $x \in \mathfrak{g}_v^{\times}$, and $x - 1 \in \mathfrak{f}_v$; if $\mathfrak{f}'$ is another integral ideal with this property, then $\mathfrak{f} \supset \mathfrak{f}'$. The ideal $\mathfrak{f}$ is called the **conductor** of ψ. Given a fractional ideal $\mathfrak{a}$ prime to $\mathfrak{f}$, we take an element α of $F_{\mathbf{h}}^{\times}$ so that $\alpha\mathfrak{g} = \mathfrak{a}$ and $\alpha_v = 1$ for every $v|\mathfrak{f}$. We then put $\psi^*(\mathfrak{a}) = \psi(\alpha)$. This is well defined. We put $\psi^*(\mathfrak{a}) = 0$ if $\mathfrak{a}$ is not prime to $\mathfrak{f}$. We call ψ^* the **ideal character associated with** ψ.

14.2. We put $G = GL_2(\mathbf{Q})$ and $G^1 = SL_2(\mathbf{Q})$ as before, and define the adelization $G_{\mathbf{A}}$ of G as usual; see [S71, §6.4] or [S97, Section 8]. We denote by $G_{\mathbf{a}}$ and $G_{\mathbf{h}}$ the archimedean and nonarchimedean factors of $G_{\mathbf{A}}$. We put $G_p = GL_2(\mathbf{Q}_p)$ for each prime number p and view G and G_p as subgroups of $G_{\mathbf{A}}$. We then put

(14.1a) $$G_{\mathbf{a}+} = \{x \in G_{\mathbf{a}} \mid \det(x) > 0\}, \quad G_{\mathbf{A}+} = G_{\mathbf{a}+}G_{\mathbf{h}}, \quad G_+ = G \cap G_{\mathbf{A}+},$$

(14.1b) $$U = G_{\mathbf{a}+}\prod_p GL(\mathbf{Z}_p), \quad U_n = \{u \in U \mid u - 1 \prec n\mathbf{Z}\},$$

(14.1c) $$U^0(n) = \{u \in U \mid b_u \prec n\mathbf{Z}\}, \quad U_0(n) = \{u \in U \mid c_u \prec n\mathbf{Z}\} \quad (0 < n \in \mathbf{Z}).$$

Here b_u and c_u are the b-entry and c-entry of u; see §1.1. Notice that $G \cap U = SL_2(\mathbf{Z}) = \Gamma(1)$ and $G \cap U_n = \Gamma(n)$. Thus $G \cap S = G^1 \cap S$ for any subgroup S of U. We note here basic facts:

(14.1d) $G^1_{\mathbf{A}} = G^1X$ and $G_{\mathbf{A}+} = G_+X \cdot \{\mathrm{diag}[1, t] \mid t \in \mathbf{Q}^\times_{\mathbf{a}}\prod_p \mathbf{Z}^\times_p,\ t_{\mathbf{a}} > 0\}$ for every open subgroup X of $G^1_{\mathbf{A}}$.

The first equality is *strong approximation* in G^1; see [S71, Lemma 6.15]. To prove the second equality, let $w \in G_{\mathbf{A}+}$ and $y = \det(w)$. We can find $a \in \mathbf{Q}^\times,\ > 0$, so that $y\mathbf{Z} = a\mathbf{Z}$. Put $t = a^{-1}y$ and $\alpha = \mathrm{diag}[1, a]$. Then $\alpha^{-1}w{\cdot}\mathrm{diag}[1, t]^{-1} \in G^1_{\mathbf{A}}$. Since $G^1_{\mathbf{A}} = G^1X$, we can put $\alpha^{-1}w \cdot \mathrm{diag}[1, t]^{-1} = \beta x$ with $\beta \in G^1$ and $x \in X$. Then $w = \alpha\beta x \cdot \mathrm{diag}[1, t]$, which proves the second equality of (14.1d).

Let $k_N = \mathbf{Q}(\mathbf{e}(1/N))$. Defining the symbol $\mathscr{A}_k$ as in §1.4, we see that $\mathscr{A}_0(\mathbf{Q}_{\mathrm{ab}}) = \bigcup_{N=1}^\infty \mathscr{A}_0(k_N, \Gamma(N))$. There is a natural action of $\mathrm{Gal}(\mathbf{Q}_{\mathrm{ab}}/\mathbf{Q})$ on the elements f of the graded algebra $\sum_{k=0}^\infty \mathscr{A}_k(\mathbf{Q}_{\mathrm{ab}})$, which can be defined as follows. For $g = \sum_{r\in\mathbf{Q}} c_r\mathbf{e}(rz) \in \mathscr{M}_k(\mathbf{Q}_{\mathrm{ab}})$ and $\sigma \in \mathrm{Gal}(\mathbf{Q}_{\mathrm{ab}}/\mathbf{Q})$ we can define $g^\sigma \in \mathscr{M}_k(\mathbf{Q}_{\mathrm{ab}})$ by $g^\sigma(z) = \sum_{r\in\mathbf{Q}} c_r^\sigma\mathbf{e}(rz)$; see Theorem 1.9 (ii). Now, for $f = g/h \in \mathscr{A}_k(\mathbf{Q}_{\mathrm{ab}})$ with $g \in \mathscr{M}_{k+m}(\mathbf{Q}_{\mathrm{ab}})$ and $0 \neq h \in \mathscr{M}_m(\mathbf{Q}_{\mathrm{ab}})$ we put $f^\sigma = g^\sigma/h^\sigma$. This is a well-defined element of $\mathscr{A}_k(\mathbf{Q}_{\mathrm{ab}})$. If $\sigma = \{t\}$ with $t \in \mathbf{Q}^\times_{\mathbf{A}}$, we write $f^{\{t\}}$ for f^σ.

In [S71, §6.6] we defined the action of $G_{\mathbf{A}+}$ on the field $\mathscr{A}_0(\mathbf{Q}_{\mathrm{ab}})$ which was written $\mathfrak{F}$ there. In [S78a] and [S78c] this was extended to an action of $G_{\mathbf{A}+}$ on the graded algebra $\sum_{k=0}^\infty \mathscr{A}_k(\mathbf{Q}_{\mathrm{ab}})$ as a group of automorphisms as follows. For $f \in \sum_{k=0}^\infty \mathscr{A}_k(\mathbf{Q}_{\mathrm{ab}})$ and $x \in G_{\mathbf{A}+}$ we denote by $f^{[x]}$ the image of f under x. As stated in [S78a, Theorem 1.2] and [S78c, Theorem 1.5], the actions have the following properties:

(14.2a) $$f^{[xy]} = (f^{[x]})^{[y]}.$$

(14.2b) $$a^{[x]} = a^{\{\det(x)^{-1}\}} \text{ for every } a \in \mathbf{Q}_{\mathrm{ab}} \text{ and } x \in G_{\mathbf{A}+}.$$

(14.2c) $$f^{[x]} = f^{\{t\}} \text{ if } x = \mathrm{diag}[1, t^{-1}] \text{ with } t \in \textstyle\prod_p \mathbf{Z}^\times_p.$$

(14.2d) $$f^{[\alpha]}(z) = (cz+d)^{-k}f(\alpha(z)) \text{ if } \alpha = \begin{bmatrix} a & b \\ c & d \end{bmatrix} \in G_+ \text{ and } f \in \mathscr{A}_k(\mathbf{Q}_{\mathrm{ab}}).$$

(14.2e) The action of $G_{\mathbf{A}+}$ on $\mathscr{A}_0(\mathbf{Q}_{\mathrm{ab}})$ is the same as that of [S71, §6.6].

(14.2f) $\mathscr{M}_k(\mathbf{Q}_{\mathrm{ab}})$ is stable under $[x]$ for every $x \in G_{\mathbf{A}+}$.

(14.2g) For every f, the group $\{x \in G_{\mathbf{A}+} \mid f^{[x]} = f\}$ contains U_n for some n.

The last statement implies that the action of $G_{\mathbf{a}+}$ is trivial, as $G_{\mathbf{a}+} \subset U_n$. Notice that (14.2d) implies that $f^{[\alpha]} = f\|_k\alpha$ if $\alpha \in G^1$. In view of (14.1d), once we require conditions (14.2c, d, e, g), then there is a unique way to define the action of $G_{\mathbf{A}+}$ on $\mathscr{A}_k(\mathbf{Q}_{\mathrm{ab}})$. However, it is highly nontrivial to prove (14.2a) for the defined action. We will present the proof in Section A5 of the Appendix. We note also that these can be generalized to the case of automorphic forms in the higher-dimensional cases; see [S98, 26.8 and 26.10] and [S00, 10.10, 10.2] in addition to [S78a] and [S78c].

Define $F_k(z;\, a,\, b)$ as in §10.3. Given $w \in \mathbf{Q}^2/\mathbf{Z}^2$, we take an element $(a,\, b)$ of $\mathbf{Q}^2$ that represents w, and define $F_k(z, w)$ to be $F_k(z;\, a,\, b)$. Since $\mathbf{Q}^2/\mathbf{Z}^2$ is canonically isomorphic to $\prod_p \mathbf{Q}_p^2/\mathbf{Z}_p^2$, for $w \in \mathbf{Q}^2/\mathbf{Z}^2$ we can speak of its p-component w_p, which is an element of $\mathbf{Q}_p^2/\mathbf{Z}_p^2$. Then for $u \in U$ we define wu to be the element of $\mathbf{Q}^2/\mathbf{Z}^2$ such that $(wu)_p = w_p u_p$. (Here we ignore the archimedean component of u.) Now we have

(14.3) $\quad F_k(z,\, w)^{[u]} = F_k(z,\, wu)$ for every $w \in \mathbf{Q}^2/\mathbf{Z}^2$ and $u \in U$.

To prove this, we first note that

(14.3a) $\quad F_k(z,\, w)^{\{t^{-1}\}} = F_k(z,\, wx)$ if $x = \mathrm{diag}[1,\, t]$ with $t \in \prod_p \mathbf{Z}_p^\times$.

This follows from the Fourier expansions of (9.4) and (10.12) combined with our definition of the action of $\{t^{-1}\}$. Here we recall a basic fact: viewing $\mathbf{e}$ as a function on $\mathbf{Q}/\mathbf{Z} = \prod_p(\mathbf{Q}_p/\mathbf{Z}_p)$, we have

(14.4) $\quad \mathbf{e}(c)^{\{t\}} = \mathbf{e}(t^{-1}c)$ for $c \in \prod_p(\mathbf{Q}_p/\mathbf{Z}_p)$ and $t \in \prod_p \mathbf{Z}_p^\times$.

To prove (14.3), fix such w and u; take n so that $F_k(z,\, w)$ is invariant under U_n and $nw = 0$. Let $t = \det(u)$ and $x = \mathrm{diag}[1,\, t]$. Then $ux^{-1} \in G_{\mathbf{A}}^1 \subset U_n G^1$ by (14.1d). Thus $ux^{-1} \in U_n\gamma$ with $\gamma \in G^1$. Clearly $\gamma \in SL_2(\mathbf{Z})$. Thus $F_k(z,\, w)^{[u]} = F_k(z,\, w)^{[\gamma x]}$. By (14.2d), $F_k(z,\, w)^{[\gamma]} = F_k(z,\, w)\|_k\gamma$, which equals $F_k(z,\, w\gamma)$ by (9.3), (10.10b, d), and (10.13). Therefore by (14.2c) and (14.3), $F_k(z,\, w\gamma)^{[x]} = F_k(z,\, w\gamma x) = F_k(z,\, wu)$. This proves (14.3).

Lemma 14.3. (i) *Let f be an element of $\mathscr{A}_k(\mathbf{Q}_{\mathrm{ab}})$ such that $f\|_k\gamma = \chi(a_\gamma)f$ for every $\gamma \in \Gamma^0(m) \cap \Gamma_0(n)$ with positive integers m and n, where χ is a character of $(\mathbf{Z}/mn\mathbf{Z})^\times$ such that $\chi(-1) = (-1)^k$. Then $f^{[x]} = \big(\chi(a_x)f\big)^{\{\det(x)^{-1}\}}$ for every $x \in U^0(m) \cap U_0(n)$, where we view χ as a character of $\prod_{p|mn} \mathbf{Z}_p^\times$ in an obvious way.*

(ii) *Let S be a subgroup of U such that* $\mathrm{diag}[1,\, \det(x)] \in S$ *for every $x \in S$ and let $\Gamma = G \cap S$. Then $f^{[x]} = f^{\{\det(x)^{-1}\}}$ for every $x \in S$ and every $f \in \mathscr{A}_k(\mathbf{Q}_{\mathrm{ab}},\, \Gamma)$.*

PROOF. Given f as in (i), take a multiple h of mn so that $f^{[u]} = f$ for every $u \in U_h$. Given $x \in U^0(m) \cap U_0(n)$, put $t = \det(x)$ and $y = \mathrm{diag}[1, t]$. By (14.1d) we can put $xy^{-1} = u\alpha$ with $\alpha \in G^1$ and $u \in U_h$. Then $\alpha \in G^1 \cap U^0(m) \cap U_0(n) = \Gamma^0(m) \cap \Gamma_0(n)$, and so $f^{[x]} = f^{[\alpha y]} = (f\|_k\alpha)^{[y]} = \bigl(\chi(a_\alpha)f\bigr)^{[y]}$. Now $a_x - a_\alpha \prec mn\mathbf{Z}$, and so $\chi(a_\alpha) = \chi(a_x)$, which proves (i). As for (ii), take again h so that $f^{[u]} = f$ for every $u \in U_h$ and $U_h \subset S$. Given $x \in S$, put $t = \det(x)$ and $y = \mathrm{diag}[1, t]$. We again have $x = u\alpha y$ with $\alpha \in G^1$ and $u \in U_h$. Then $\alpha \in S \cap G^1 = \Gamma$, and $f^{[x]} = (f\|_k\alpha)^{[y]} = f^{[y]} = f^{\{t^{-1}\}}$, which proves (ii).

14.4. Let K be an imaginary quadratic field embedded in $\mathbf{C}$ as in §12.1 and $\mathfrak{r}$ the maximal order in K. We first recall a basic fact that $f(\tau) \in K_{\mathrm{ab}}$ for every $\tau \in K \cap H$ and every $f \in \mathfrak{F}$. (In fact, we stated in Theorem 12.2 more general results concerning the values of nearly holomorphic forms.) Fixing τ, define $q : K_{\mathbf{A}} \to M_2(\mathbf{Q}_{\mathbf{A}})$ by

$$\begin{bmatrix}\tau\\1\end{bmatrix} s = q(s)\begin{bmatrix}\tau\\1\end{bmatrix} \qquad (s \in K_{\mathbf{A}}). \tag{14.5}$$

Then the reciprocity-law [S71, Theorem 6.31] says

(14.6) $f(\tau)^{\{s\}} = f^{[q(s)]^{-1}}(\tau)$ for every $s \in K_{\mathbf{A}}^\times$ and every $f \in \mathfrak{F}$ finite at τ.

We now consider an elliptic curve C which is isomorphic to $\mathbf{C}/\mathfrak{a}$ and rational over an algebraic number field k, where $\mathfrak{a}$ is a $\mathbf{Z}$-lattice in K. We fix an isomorphism $\xi : \mathbf{C}/\mathfrak{a} \to C$. For $b \in K$ such that $b\mathfrak{a} \subset \mathfrak{a}$ we denote by $\iota(b)$ the element of $\mathrm{End}(C)$ obtained from multiplication by b on $\mathbf{C}$, that is, $\iota(b)\xi(w) = \xi(bw)$ for $w \in \mathbf{C}/\mathfrak{a}$. We can naturally extend ι to an isomorphism of K onto $\mathrm{End}(C) \otimes \mathbf{Q}$. For a holomorphic 1-form η on C we have $\eta \circ \iota(b) = b\eta$. (Indeed, if u is the variable on $\mathbf{C}$, then η is a constant times du.) We can take η to be k-rational. From the equality $\eta \circ \iota(b) = b\eta$ we see that every element of $\mathrm{End}(C)$ is rational over kK.

Let us next recall some basic results concerning the zeta function of C over k, which is associated with a Hecke character of $k_{\mathbf{A}}^\times$. We first assume that $K \subset k$. For $z \in k_{\mathbf{A}}^\times$ we denote by $z\mathfrak{a}$ the $\mathbf{Z}$-lattice in K such that $(z\mathfrak{a})_p = z_p\mathfrak{a}_p$ for every prime p, where $z_p = (z_v)_{v|p}$. The module $K/\mathfrak{a}$ can be identified with $\prod_p(K_p/\mathfrak{a}_p)$. Therefore multiplication by z_p on $K_p/\mathfrak{a}_p$ defines an isomorphism of $K/\mathfrak{a}$ onto $K/z\mathfrak{a}$. In particular, if $z\mathfrak{a} = \mathfrak{a}$, multiplication by z in this sense defines an automorphism of $K/\mathfrak{a}$.

Clearly $\xi(K/\mathfrak{a})$ is the set of all torsion points of C. Moreover, it can be shown that all such points are rational over k_{ab}; see [S71, Proposition 7.40 (1)]. Therefore, for $x \in k_{\mathbf{A}}^\times$ and $w \in K/\mathfrak{a}$, the image $\xi(w)^{\{x\}}$ of $\xi(w)$ under $\{x\}$ is meaningful. Put $y = N_{k/K}(x)$. Then there exists a homomorphism $\alpha : k_{\mathbf{A}}^\times \to K^\times$ with the following properties:

(14.7a) $$\alpha(x)\mathfrak{a} = y\mathfrak{a}, \quad \alpha(x)\alpha(x)^\rho = N(y\mathfrak{r}),$$

(14.7b) $$\xi(w)^{\{x\}} = \xi\big(\alpha(x)y^{-1}w\big) \text{ for } w \in K/\mathfrak{a}.$$

This result is included in [S71, Proposition 7.40]; see also [S98, 19.8 and 19.10]. We can then define a Hecke character ψ of $k_{\mathbf{A}}^\times$ by

(14.8) $$\psi(x) = |y|_{\mathbf{A}}^{1/2}\big(\alpha(x)y^{-1}\big)_{\mathbf{a}}, \quad y = N_{k/K}(x).$$

Since $K_{\mathbf{a}}$ is identified with $\mathbf{C}$, we have $z_{\mathbf{a}} \in \mathbf{C}$ for every $z \in K_{\mathbf{A}}^\times$. Notice that

(14.8a) $$\psi(x) = y^{-1}|y| \text{ if } x \in k_{\mathbf{a}}^\times \text{ and } y = N_{k/K}(x).$$

The character ψ in [S71] and [S98] is defined without the factor $|y|_{\mathbf{A}}^{1/2}$. Here we include it in conformity with the definition in §13.5. Let $\mathfrak{g}$ denote the maximal order of k. Since $|N_{k/K}(x)|_{\mathbf{A}} = N(x\mathfrak{g})^{-1}$, (14.8) shows that

(14.8b) $$\psi(x)N(x\mathfrak{g})^{1/2} = \alpha(x) \in K^\times \text{ if } x \in K_{\mathbf{h}}^\times.$$

Thus (14.7a) and (14.8a, b) are necessary conditions for ψ. Conversely given ψ, we can find C as follows. Let $\mathfrak{a}$ be a $\mathbf{Z}$-lattice in K and k a finite algebraic extension of $K(j)$, where j is the invariant of $\mathbf{C}/\mathfrak{a}$. Let ψ be a Hecke character of $k_{\mathbf{A}}^\times$ such that (14.7a) and (14.8a, b) hold with a map $\alpha : k_{\mathbf{A}}^\times \to K^\times$. Then there exists a k-rational elliptic curve C and a parametrization $\xi : \mathbf{C}/\mathfrak{a} \to C$ satisfying (14.7b); see [S98, 22.1].

Let us now state this fact in terms of an ideal character. Let ψ^* be the ideal character associated with ψ. Let $\mathfrak{f}$ be the conductor of ψ. Then we can define a character $\psi_{\mathfrak{f}}$ of $(\mathfrak{g}/\mathfrak{f})^\times$ by $\psi_{\mathfrak{f}}(a) = \prod_{v|\mathfrak{f}}\psi_v(a)$. Put $\alpha(\mathfrak{x}) = \psi^*(\mathfrak{x})N(\mathfrak{x})^{1/2}$. Then (14.8b) shows that $\alpha(\mathfrak{x}) \in K^\times$. In particular, if $\gamma \in k^\times$ and γ is prime to $\mathfrak{f}$, then

(14.8c) $$\psi^*(\gamma\mathfrak{g}) = \psi_{\mathfrak{f}}(\gamma)^{-1}\psi_{\mathbf{a}}(\gamma)^{-1} = \psi_{\mathfrak{f}}(\gamma)^{-1}N_{k/K}(\gamma)|N_{k/K}(\gamma)|^{-1},$$

and so $\alpha(\gamma\mathfrak{g}) = \psi_{\mathfrak{f}}(\gamma)^{-1}N_{k/K}(\gamma)$.

Now the zeta function of C over k is $L(s, \psi)L(s, \overline{\psi})$; see [S71, Theorem 7.43] and [S98, 19.11]. We call ψ the **Hecke character determined by C over k.** It can easily be verified that this does not depend on the choice of $\mathfrak{a}$ and ξ.

We state two basic facts on the zeta function of C in the following lemma.

Lemma 14.5. (i) *Let C and C' be two elliptic curves which are isomorphic over $\mathbf{C}$ and defined over a finite algebraic extension k of K. If they determine the same Hecke character over k, then they are isomorphic over k.*

(ii) *Suppose C is defined over a subfield h of k such that $k = hK$ and $K \not\subset h$. If ψ is the Hecke character of $k_{\mathbf{A}}^\times$ determined by C over k, then $L(s, \psi)$ is the zeta function of C over h.*

Assertion (i) is a special case of [S98, 19.12]; as for (ii), see[S98, 20.7 and 22.2]. We must note, however, that there can exist C and C' as in (i) which are defined over a field h as in (ii), but not isomorphic over h. We will discuss this point in more detail in Section 15B.

14.6. There is an interesting special situation in which the above character is given in the form $\psi = \mu \circ N_{k/K}$ with a Hecke character μ of $K_{\mathbf{A}}^{\times}$. (We are assuming that $K \subset k$.) Such a μ exists if and only if every point of $\xi(K/\mathfrak{a})$ is rational over $K_{\text{ab}}k$, and moreover the number of such μ is exactly $[k \cap K_{\text{ab}} : K]$; see [S71, Theorem 7.44]. We can state the result in a clear-cut way under the following condition:

(14.9) *Every point of* $\xi(K/\mathfrak{a})$ *is rational over* K_{ab}.

In this case C is clearly rational over K_{ab}, and for $w \in K/\mathfrak{a}$ and $\sigma \in \text{Gal}(K_{\text{ab}}/K)$ the image $\xi(w)^{\sigma}$ is meaningful as a point of C^{σ}.

Theorem 14.7. *Under (14.9) let* k *be a finite extension of* K *contained in* K_{ab} *such that* C *is* k*-rational. Then there exists a homomorphism* $\beta : K^{\times}N_{k/K}(k_{\mathbf{A}}^{\times}) \to K^{\times}$ *uniquely determined by the following properties:*

$$\beta(y)\mathfrak{a} = y\mathfrak{a}, \quad \beta(y)\beta(y)^{\rho} = N(y\mathfrak{r}), \tag{14.10a}$$

$$\xi(w)^{\{y\}} = \xi\big(\beta(y)y^{-1}w\big) \ \text{for every} \ w \in K/\mathfrak{a}, \tag{14.10b}$$

where y *is the variable element of* $K^{\times}N_{k/K}(k_{\mathbf{A}}^{\times})$. *Moreover, there exists a Hecke character* μ *of* $K_{\mathbf{A}}^{\times}$ *such that*

$$\mu(y) = |y|_{\mathbf{A}}^{1/2}\big(\beta(y)y^{-1}\big)_{\mathbf{a}} \ \text{for} \ y \in K^{\times}N_{k/K}(k_{\mathbf{A}}^{\times}), \tag{14.11}$$

and $\mu \circ N_{k/K}$ *is the Hecke character of* $k_{\mathbf{A}}^{\times}$ *determined by* C *over* k.

PROOF. By class field theory, k corresponds to $K^{\times}N_{k/K}(k_{\mathbf{A}}^{\times})$, and for $y \in K_{\mathbf{A}}^{\times}$ the element $\{y\}$ of $\text{Gal}(K_{\text{ab}}/K)$ is the identity map on k if and only if $y \in K^{\times}N_{k/K}(k_{\mathbf{A}}^{\times})$. Thus $\xi(w)^{\{y\}}$ belongs to C. Given $y \in K^{\times}N_{k/K}(k_{\mathbf{A}}^{\times})$, take $x \in k_{\mathbf{A}}^{\times}$ and $y_1 \in K^{\times}$ so that $y = y_1N_{k/K}(x)$. Putting $\beta(y) = y_1\alpha(x)$, from (14.7a, b) we obtain (14.10a, b), and we easily see that $\mu \circ N_{k/K}$ coincides with ψ of (14.8). Formula (14.11) defines μ only on $K^{\times}N_{k/K}(k_{\mathbf{A}}^{\times})$. Since $[K_{\mathbf{A}}^{\times} : K^{\times}N_{k/K}(k_{\mathbf{A}}^{\times})] = [k : K] < \infty$, μ can be extended to a continuous character of $k_{\mathbf{A}}^{\times}$ by a basic principle stated in [S71, Lemma 7,45].

14.8. We now take C to be a member of a family of elliptic curves parametrized by $z \in H$ as follows. Take $\varphi \in \mathscr{A}_1(\mathbf{Q}_{\text{ab}})$, and put $\omega_1 = z\omega_2$ and $\omega_2 = 2\pi i\varphi(z)$. Then $g_{\mu}(\omega_1, \omega_2) = (2\pi i)^{-2\mu}g_{\mu}(z, 1)/\varphi(z)^{2\mu} = \mathbf{g}_{\mu}(z)/\varphi(z)^{2\mu}$ for $\mu = 2, 3$. Once φ is fixed, we have a family of elliptic curves (viewed as projective curves in an obvious way)

$$C_z : Y^2 = 4X^3 - c_2(z)X - c_3(z), \quad c_{\mu}(z) = \mathbf{g}_{\mu}(z)/\varphi(z)^{2\mu}. \tag{14.12}$$

This is what we denoted by $C(z\omega, \omega)$ in (10.4e), where $\omega = 2\pi i\varphi(z)$. The curve C_z is meaningful if $\varphi(z) \neq 0, \infty$. Notice that c_2 and c_3 are finite at z if and only if $\varphi(z) \neq 0$, as $c_2^3 - 27c_3^2 = \Delta/\varphi^{12}$. Now for $\tau \in K \cap H$ such that $\varphi(\tau) \neq 0, \infty$, the curve C_τ is isomorphic to $\mathbf{C}/(\mathbf{Z}\tau + \mathbf{Z})$. We see that $c_\mu \in \mathscr{A}_0(\mathbf{Q}_{\mathrm{ab}})$, and so $c_\mu(\tau) \in K_{\mathrm{ab}}$, as we noted in §12.1. Thus C_τ is K_{ab}-rational. The significance of this curve C_τ is that we have an explicit form of the period $\omega_2 = 2\pi i\varphi(\tau)$ and we can determine the Hecke character associated with C_τ, as we will show in Theorem 14.9 below. Thus the critical value of the zeta function of C_τ can be compared with $\varphi(\tau)$ by the technique of Section 13.

Put $\mathfrak{a} = \mathbf{Z}\tau + \mathbf{Z}$. Then the map $u \mapsto u\omega_2$ gives an isomorphism of $\mathbf{C}/\mathfrak{a}$ onto $\mathbf{C}/(\mathbf{Z}\omega_1 + \mathbf{Z}\omega_2)$. For $w = a\tau + b \in K$ with $a, b \in \mathbf{Q}$, let $\xi(w)$ denote the point (X, Y) on C_τ corresponding to w. More explicitly, from (10.6a), (10.10a), and (10.11a) we see that

$$(14.13)\qquad \begin{aligned}\xi(w) &= \big(\wp(a\omega_1 + b\omega_2;\, \omega_1,\, \omega_2),\ \wp'(a\omega_1 + b\omega_2;\, \omega_1,\, \omega_2)\big)\\ &= \big(\varphi(\tau)^{-2}F_2(\tau; a,\, b),\ \varphi(\tau)^{-3}F_3(\tau; a,\, b)\big).\end{aligned}$$

Thus $\xi(w)$ is K_{ab}-rational for every $w \in K$, and so (14.9) is satisfied.

Theorem 14.9. *Define φ and C_z as above; let τ be an element of $K \cap H$ such that $\varphi(\tau) \neq 0$; define also $q : K_{\mathbf{A}}^\times \to G_{\mathbf{A}+}$ by (14.5). Put $U_\varphi = \{u \in U \mid \varphi^{[u]} = \varphi\}$ and $W = \{y \in K_{\mathbf{A}}^\times \mid q(y) \in U_\varphi\}$. Let k be a subfield of K_{ab} corresponding to $K^\times W$. (The field k is meaningful, as W is an open subgroup of $K_{\mathbf{A}}^\times$, and $K^\times W = K^\times N_{k/K}(k_{\mathbf{A}}^\times)$.) Then $c_2(\tau), c_3(\tau) \in k$, and consequently C_τ is k-rational. Moreover, for $y \in K^\times W$ the quantity $\beta(y)$ of (14.10b) is a unique element of $K^\times$ such that $y \in \beta(y)W$.*

PROOF. By Lemma 14.3, $\mathbf{g}_\nu^{[u]} = \mathbf{g}_\nu$ for $u \in U$, and so $c_\nu^{[u]} = c_\nu$ for $u \in U_\varphi$. For $y \in W$ put $x = q(y)^{-1}$. Then $x \in U_\varphi$, and by (14.6), $c_\nu(\tau)^{\{y\}} = c_\nu^{[x]}(\tau) = c_\nu(\tau)$. Therefore $c_\nu(\tau)$ belongs to the field k of our theorem, and so C_τ is k-rational. For $y \in K^\times W$ put $y = ba$ with $b \in K^\times$ and $a \in W$; put also $r = q(a)^{-1}$. For $w = c\tau + d \in K/\mathfrak{a}$ with $(c, d) \in \mathbf{Q}^2/\mathbf{Z}^2$, we see from (14.5) that $a^{-1}w = c'\tau + d'$ with $(c', d') = (c, d)r$. Since $\{y\} = \{a\}$ and $\varphi^{[r]} = \varphi$, by (14.6) and (14.3) we obtain

$$\{\varphi(\tau)^{-\nu}F_\nu(\tau;\, c,\, d)\}^{\{y\}} = \{\varphi(\tau)^{-\nu}F_\nu(\tau;\, c,\, d)\}^{\{a\}} = \varphi(\tau)^{-\nu}F_\nu(\tau;\, c',\, d').$$

Therefore, by (14.13) we have $\xi(w)^{\{y\}} = \xi(a^{-1}w) = \xi(by^{-1}w)$ for every $w \in K/\mathfrak{a}$. This means that $b = \beta(y)$, which proves our theorem.

14.10. There is an important requirement in the setting of §14.6, that is, we have to choose φ so that $\varphi(\tau) \neq 0$. In certain cases we can check this point in the following way. Take, for example, the group Γ conjugate to $\Gamma_1(4)$ treated in §10.7. Let $0 \neq \varphi \in \mathscr{M}_1(\Gamma)$. The divisor of φ, written $\mathrm{div}(\varphi)$, can be defined as in [S71, §2.4], and given by the formula [S71, (2.6.5)], and so $\deg\big(\mathrm{div}(\varphi)\big)$ is computable. As noted in §1.11, there are two regular cusps and one irregular cusp. Thus we find that $\deg\bigl(\mathrm{div}(\varphi)\bigr) = 1/2$. Since the order of a zero of φ at any irregular cusp is $1/2$ times an odd integer, we see that φ is not zero at all other points of $H \cup \mathbf{Q} \cup \{\infty\}$.

As another example, take

$$\Gamma = \{\gamma \in \Gamma_0(8) \mid \chi(d_\gamma) = 1\}, \quad \chi(a) = \left(\frac{-8}{a}\right).$$

Then $\Gamma_0(8) = \{\pm 1\}\Gamma$. This has genus 0 and no elliptic points; there are two regular and two irregular cusps, as shown at the end of Section 1. We find that $\dim \mathscr{M}_k(\Gamma) = 1 + 2[k/2]$ for every $k \geq 0$. Define $\varphi^{\ell}_{k,\chi}$ and $\varphi^{r}_{k,\chi}$ by (9.10a, b) with the present χ and $N = 8$. Put $\varphi = \varphi^{r}_{1,\chi}$. Then $\varphi(z) = \varphi^{\ell}_{1,\chi}(8z)$ by (9.11h). Again from [S71, (2.6.5)] we obtain $\deg\bigl(\mathrm{div}(\varphi)\bigr) = 1$. Therefore, for the same reason as in the previous case, we see that φ has 0 exactly at the irregular cusps, and $\varphi \neq 0$ at all other points of $H \cup \mathbf{Q} \cup \{\infty\}$.

14.11. We note that if we put $S = \{x \in G_{\mathbf{A}+} \mid c_2^{[x]} = c_2,\ c_3^{[x]} = c_3\}$, then

(14.14) $$S = \mathbf{Q}^\times U_\varphi.$$

Indeed, in the above proof we have seen that $U_\varphi \subset S$, and so $\mathbf{Q}^\times U_\varphi \subset S$. To prove $S \subset \mathbf{Q}^\times U_\varphi$, we first recall

(14.14a) $$\mathbf{Q}^\times U_n = \{x \in G_{\mathbf{A}+} \mid f^{[x]} = f \text{ for every } f \in \mathscr{A}_0\bigl(k_n, \Gamma(n)\bigr)\},$$

as given in [S71, (6.6.3)]. Let $x \in S$. Since J is a rational expression in c_2 and c_3 with rational coefficients, we have $J^{[x]} = J$, and so $x \in \mathbf{Q}^\times U$, as $\mathbf{Q}(J) = \mathscr{A}_0\bigl(\mathbf{Q}, \Gamma(1)\bigr)$. Put $x = au$ with $a \in \mathbf{Q}^\times$ and $u \in U$. Since $\mathbf{g}_\nu^{[u]} = \mathbf{g}_\nu$, we have $(\varphi^2)^{[u]} = \varphi^2$, and so $\varphi^{[u]} = \pm\varphi$. Clearly $\varphi^{[-1]} = -\varphi$. Thus $u \in \{\pm 1\}U_\varphi$, which proves (14.14).

Let k_S be the subfield of $\mathbf{Q}_{\mathrm{ab}}$ corresponding to $\mathbf{Q}^\times \det(S)$ and let

(14.15) $$\mathfrak{F}_S = \{f \in \mathscr{A}_0(\mathbf{Q}_{\mathrm{ab}}) \mid f^{[x]} = f \text{ for every } x \in S\}.$$

Then we easily see that $k_S = \mathfrak{F}_S \cap \mathbf{Q}_{\mathrm{ab}}$ and $k_S(c_2, c_3) \subset \mathfrak{F}_S$. We have actually $\mathfrak{F}_S = k_S(c_2, c_3)$ in view of the fact that the correspondence $S \leftrightarrow \mathfrak{F}_S$ is bijective; see [S71, (6.7.3)]. We have

(14.15a) $$k_S = \mathbf{Q} \quad \text{if} \quad \varphi \in \mathscr{A}_1(\mathbf{Q}).$$

Indeed, (14.2c) shows that $\mathrm{diag}[1, t] \in U_\varphi$ for every $t \in \prod_p \mathbf{Z}_p^\times$ if $\varphi \in \mathscr{A}_1(\mathbf{Q})$, and so $\mathbf{Q}_{\mathbf{A}}^\times \subset \mathbf{Q}^\times \det(S)$, which proves (14.15a).

It is natural to ask whether the field k of Theorem 14.9 is $k_S K\bigl(c_2(\tau), c_3(\tau)\bigr)$. We can at least prove a weaker result:

(14.16) $$k_S K\bigl(c_2(\tau), c_3(\tau)\bigr) \subset k.$$

Indeed, for $y \in W$ we have $yy^\rho = \det\bigl(q(y)\bigr) \in \det(S)$, and so for $a \in k_S$ we have $a^{\{y\}} = a$. Thus $k_S \subset k$, and so (14.16) holds. It is not clear, however, that the inclusion is indeed an equality. We can actually prove a stronger result $k = k_S K\bigl(c_2(\tau), c_3(\tau)\bigr)$ *under the following condition:*

$$\text{(14.17)} \qquad \gamma \in \Gamma(1) \text{ and } (\varphi\|_1\gamma)(\tau) = \varphi(\tau) \implies \gamma \in U_\varphi.$$

Indeed, take $y \in K_{\mathbf{A}}^\times$ and suppose that $\{y\}$ is the identity map on the field $k_S\bigl(c_2(\tau), c_3(\tau)\bigr)$; put $r = q(y)^{-1}$. Then $\det(r)^{-1} = yy^\rho$, and $\{\det(r)\}$ is the identity map on k_S, so that $\det(r) \in \mathbf{Q}^\times \det(U_\varphi)$. Put $\det(r) = c\det(x)$ with $c \in \mathbf{Q}^\times$ and $x \in U_\varphi$. Put $c = \det(\xi)$ with $\xi \in G$. Then $\det(x^{-1}r\xi^{-1}) = 1$. By (14.1d) we can put $x^{-1}r\xi^{-1} = u\eta$ with $u \in U_\varphi$ and $\eta \in G^1$. Put $\alpha = \eta\xi$ and $v = xu$. Then $r = v\alpha$ and $v \in U_\varphi$. Since r and v belong to $G_{\mathbf{A}+}$, we see that $\alpha \in G_+$. Now $c_\nu(\tau) = c_\nu(\tau)^{\{y\}} = c_\nu^{[r]}(\tau)$ and $c_\nu^{[r]} = c_\nu^{[v\alpha]} = c_\nu^{[\alpha]} = c_\nu \circ \alpha$. Thus $c_\nu(\tau) = c_\nu(\alpha\tau)$, and so $J(\tau) = J(\alpha\tau)$. Therefore $\alpha\tau = \gamma\tau$ with $\gamma \in \Gamma(1)$. Since $q(K^\times) = \bigl\{\varepsilon \in G_+ \bigm| \varepsilon\tau = \tau\bigr\}$, we have $\gamma^{-1}\alpha = q(b)$ with $b \in K^\times$. Since $c_\nu(\tau) = c_\nu(\gamma\tau)$, we have $(\varphi\|_1\gamma)^{2\nu}(\tau) = \varphi(\tau)^{2\nu}$, provided $c_\nu(\tau) \neq 0$. Suppose $c_2(\tau)c_3(\tau) \neq 0$. Then $(\varphi\|_1\gamma)(\tau) = \pm\varphi(\tau)$. Replacing γ by $-\gamma$ if necessary, we have $\gamma \in U_\varphi$ by virtue of our assumption (14.17). Thus $q(yb^{-1}) = r\alpha^{-1}\gamma = v\gamma \in U_\varphi$, and so $yb^{-1} \in W$. This means that $\{y\}$ is the identity map on k, and proves that $k = k_S K\bigl(c_2(\tau), c_3(\tau)\bigr)$. If $c_2(\tau) = 0$, then $c_3(\tau) \neq 0$, $[\mathfrak{r}^\times : 1] = 6$, and $(\varphi\|_1\gamma)^6(\tau) = \varphi(\tau)^6$. Thus $(\varphi\|_1\gamma)(\tau) = \zeta\varphi(\tau)$ with a sixth root of unity ζ. Since $[\mathfrak{r}^\times : 1] = 6$, we can find an element ω of $\Gamma(1)$ such that $\omega(\tau) = \tau$ and $j_\omega(\tau) = \zeta$. Then $(\varphi\|_1\gamma\omega)(\tau) = \varphi(\tau)$, which combined with (14.17) shows that $\gamma\omega \in U_\varphi$. Replacing γ by $\gamma\omega$, we obtain the desired conclusion. The case $c_3(\tau) = 0$ can be handled in a similar way.

14.12. We add here three remarks.

(1) Condition (14.17) is satisfied for every $\tau \in K \cap H$ if $\varphi = \Delta^{1/12}$. Indeed, for $\gamma \in \Gamma(1)$ and $\varphi = \Delta^{1/12}$ we have $\varphi\|_1\gamma = c_\gamma\varphi$ with $c_\gamma \in \mathbf{C}^\times$. Therefore, if $(\varphi\|_1\gamma)(\tau) = \varphi(\tau)$, then $c_\gamma = 1$, and so $\varphi\|_1\gamma = \varphi$. Thus $\gamma \in U_\varphi$ as expected. By (14.15a) we have $k_S = \mathbf{Q}$, and so $k = K\bigl(c_2(\tau), c_3(\tau)\bigr)$ for every $\tau \in K \cap H$ if $\varphi = \Delta^{1/12}$.

(2) The above theorem implies that $K^\times \cap W = \{1\}$, but the fact can be proved more directly as follows. Let $\gamma = q(\varepsilon)$ with $\varepsilon \in K^\times \cap W$. Then $\gamma \in G \cap U_\varphi \subset SL_2(\mathbf{Z})$, and $\varphi = \varphi^{[\gamma]} = \varphi_1\|\gamma$. Since $\gamma(\tau) = \tau$, we have $\varphi(\tau) = j_\gamma(\tau)\varphi(\tau)$, and so $j_\gamma(\tau) = 1$, as $\varphi(\tau) \neq 0$. Since the map $\gamma \mapsto j_\gamma(\tau)$ is injective, we have $\gamma = 1$ as expected.

(3) A result similar to the above theorem was given by Rumely [Ru]. However, there is a gap in his proof. He invoked [S71, Proposition 6.33] by taking the map $z \mapsto \bigl(c_2(z), c_3(z)\bigr)$ as the biregular map φ_S of $(G \cap S)\backslash H$ onto a nonsingular curve, but it is not clear that the image of the map is indeed nonsingular. With-

out that knowledge we cannot say that it defines a biregular map, an essential requirement of the quoted result. We can show that the image is nonsingular and the map defines a biregular map if $\varphi = \Delta^{1/12}$ as follows. In this case we have $c_2^3 - 27c_3^2 = 1$, and so the map $z \mapsto (x, y) = \bigl(c_2(z), c_3(z)\bigr)$ sends $(G \cap S)\backslash H$ onto an elliptic curve $27y^2 = x^3 - 1$, which is nonsingular, and so we can employ [S71, Proposition 6.33]. This gives another proof of the equality $k = K\bigl(c_2(\tau), c_3(\tau)\bigr)$ for every $\tau \in K \cap H$ when $\varphi = \Delta^{1/12}$. We also note that [Ru] contains the investigations of other types of elliptic curves with complex multiplication for which the corresponding Hecke characters can be determined, as well as many numerical examples.

14.13. To make our exposition smooth, we define a subgroup $\mathfrak{r}_{\mathbf{h}}^\times$ of $K_{\mathbf{A}}^\times$ by

$$\mathfrak{r}_{\mathbf{h}}^\times = \prod_p \mathfrak{r}_p^\times, \quad \mathfrak{r}_p = \mathfrak{r} \otimes_{\mathbf{Z}} \mathbf{Z}_p. \tag{14.18}$$

Now, as an easy example of Theorem 14.9 let us consider the case where $\varphi(z) = \sqrt{-3}\,F_1(z;\, 0,\, 1/3)$ and $\tau = \zeta = \mathbf{e}(1/3)$; thus $K = \mathbf{Q}(\zeta)$. From the last equality of (13.36) we see that $c_3(\zeta) = -1$, and so our curve C_ζ is $Y^2 = 4X^3 + 1$. Notice also that $\varphi \in \mathscr{M}_1(\mathbf{Q})$. We easily see that for $\gamma = \sqrt{-3}$ the map $(X, Y) \mapsto (\gamma^2 X, \gamma^3 Y)$ gives an isomorphism of C_ζ to the curve $C(\zeta\omega, \omega)$ of (13.37) over K, but C_ζ is not isomorphic to $C(\zeta\omega, \omega)$ over $\mathbf{Q}$. (If we take $\varphi(z) = F_1(z;\, 0,\, 1/3)$ instead of the above choice, then C_ζ becomes the curve of (13.37).)

Let χ be a nontrivial character of $(\mathbf{Z}/3\mathbf{Z})^\times$. We can view χ as a character of $\mathfrak{r}_{\mathbf{h}}^\times$ by combining it with the natural homomorphism $\mathfrak{r}_{\mathbf{h}}^\times \to (\mathfrak{r}/\sqrt{-3}\mathfrak{r})^\times \cong (\mathbf{Z}/3\mathbf{Z})^\times$. Since $\varphi \in \mathscr{M}_1\bigl(\Gamma_1(3), \mathbf{Q}\bigr)$, by Lemma 14.3 (i) we have $\varphi^{[x]} = \chi(a_x)\varphi$ for $x \in U_0(3)$. For $y \in \mathfrak{r}_{\mathbf{h}}^\times$ we can find an element ε of $\mathfrak{r}^\times$ such that $\varepsilon^{-1}y - 1 \prec 3\mathfrak{r}$. Define q by (14.5) with ζ in place of τ. Then we see that $q(\varepsilon^{-1}y) \in U_\varphi$, and so $y \in \varepsilon W$. Thus $\mathfrak{r}_{\mathbf{h}}^\times \subset K^\times W$, and so $K^\times W = K_{\mathbf{A}}^\times$, as $K_{\mathbf{A}}^\times = K^\times K_{\mathbf{a}}^\times \mathfrak{r}_{\mathbf{h}}^\times$ for the present K. This means that $k = K$, and moreover $\beta(y) = \varepsilon$. Therefore the Hecke character of $K_{\mathbf{A}}^\times$ determined by C_ζ is μ of (14.11). Let λ be the ideal character of K corresponding to $\overline{\mu}$. Then $\lambda(\alpha\mathfrak{r}) = \alpha^{-1}|\alpha|$ for $\alpha \in \mathfrak{r}$ such that $\alpha - 1 \in 3\mathfrak{r}$. In Lemma 14.5 (ii) we mentioned a result about the zeta function of C over a field h not containing K. In the present case we have $h = \mathbf{Q}$, and so $L(s, \lambda)$ is the zeta function of both C and C_ζ over $\mathbf{Q}$. This character λ is the same as that of §13.15 and so we showed in (13.39) that $L(1/2, \lambda) = \omega/3$ with ω of (13.38). On the other hand, the period ω_2 on C_ζ is $2\pi i\varphi(\zeta) = 2\pi i\sqrt{-3}\,F_1(\zeta;\, 0,\, 1/3)$, which equals $-\sqrt{-3}\,\omega$. Thus the quotient of the periods, $-\sqrt{-3}$, belongs to K, but not to $\mathbf{Q}$. This is a typical example of a general principle, which we will discuss in more detail in Section 15B. In the simplest case in which K has class number 1, this can be explained as follows. Given K, $\mathfrak{a}$, and a Hecke ideal character λ of K, there exists a $\mathbf{Q}$-rational elliptic curve isomorphic to $\mathbf{C}/\mathfrak{a}$ whose zeta function is $L(s, \lambda)$. All such curves are isomorphic over K, and form exactly two isomorphism classes over $\mathbf{Q}$. The quotient of $L(1/2, \lambda)$ by the period of a $\mathbf{Q}$-rational 1-form on a 1-cycle on the curve depends on the choice of the isomorphism class. Of course the period depends on the choice of the 1-cycle. We will state a precise formulation in Theorem 15.10.

14.14. To find other examples in many more cases, we take an integral ideal $\mathfrak{c}$ in K, a positive integer $m > 2$, and a character $\chi : (\mathfrak{r}/\mathfrak{c})^\times \to \{\pm 1\}$ satisfying the following condition:

(14.19) *The injection* $\mathbf{Z} \to \mathfrak{r}$ *gives a bijection of* $\mathbf{Z}/m\mathbf{Z}$ *onto* $\mathfrak{r}/\mathfrak{c}$ *and* $\chi(-1) = -1$.

Then $N(\mathfrak{c}) = m$. We can view χ as a character of $\mathfrak{r}_\mathbf{h}^\times$ by combining it with the natural homomorphism $\mathfrak{r}_\mathbf{h}^\times \to \prod_p (\mathfrak{r}_p/\mathfrak{c}_p)^\times \cong (\mathfrak{r}/\mathfrak{c})^\times$. Let φ be an element of $\mathscr{A}_1(\mathbf{Q})$ such that $\varphi\|_1\gamma = \chi(a_\gamma)\varphi$ for every $\gamma \in \Gamma^0(m)$. We then consider the curve C_z of (14.10) with this φ.

Under (14.19) we have $\mathfrak{r} = \mathbf{Z} + \mathfrak{c}$, and so by Lemma 12.6 every fractional ideal in K is equivalent to an integral ideal $\mathfrak{a}$ of the form

(14.20) $\mathfrak{a} = \mathbf{Z}\sigma + \mathbf{Z}r, \quad \sigma \in \mathfrak{c} \cap H, \quad \mathfrak{r} = \mathbf{Z}\sigma + \mathbf{Z}, \quad r = N(\mathfrak{a}), \quad r$ is prime to m,

Thus the question about $\mathbf{C}/\mathfrak{x}$ for a fractional ideal $\mathfrak{x}$ can be reduced to $\mathbf{C}/\mathfrak{a}$ with $\mathfrak{a}$ as in (14.20).

Theorem 14.15. *Let* $\mathfrak{c}$, m, χ, *and* φ *be as in* §14.14; *also let* $\mathfrak{a}$, σ, *and* r *be as in (14.20). Putting* $\tau = \sigma/r$ *and assuming that* $\varphi(\tau) \neq 0$, *define* k, U_φ, *and* W *as in Theorem 14.9. Then* $\mathfrak{r}_\mathbf{h}^\times \subset K^\times W$, k *is the Hilbert class field over* K, $k = K(c_2(\tau), c_3(\tau))$, *and* $\beta(y)$ *of (14.11) can be given by* $\beta(y) = \chi(y)$ *for* $y \in \mathfrak{r}_\mathbf{h}^\times$. *Further let* μ *be a Hecke character of* $K_\mathbf{A}^\times$ *satisfying (14.11) with respect to* C_τ, *and* λ *the ideal character corresponding to* $\overline{\mu}$. *Then the conductor of* λ *divides* $\mathfrak{c}$, *and* $\lambda(\alpha\mathfrak{r}) = \chi(\alpha)\alpha^{-1}|\alpha|$ *for every* $\alpha \in K^\times$ *prime to* $\mathfrak{c}$.

PROOF. Given $y \in \mathfrak{r}_\mathbf{h}^\times$, put $s = \chi(y)y = e\sigma + f$ with $e, f \in \mathbf{Q}_\mathbf{A}$. Then $e, f \prec \mathbf{Z}$, as $\mathfrak{r} = \mathbf{Z}\sigma + \mathbf{Z}$. Define q by (14.5) with the present τ and put $u = q(s)$. Put also $\sigma + \sigma^\rho = g$ and $\sigma\sigma^\rho/r = h$. Then $u = \begin{bmatrix} eg+f & -eh \\ er & f \end{bmatrix}$. We have $m|h$, as $\sigma \in \mathfrak{c}$, and so $u \in U^0(m)$. Now $a_u - s = eg + f - (e\sigma + f) = e\sigma^\rho \in \mathfrak{c}$, and so $\chi(a_u) = \chi(s) = 1$. Thus $\varphi^{[u]} = \varphi$ by Lemma 14.3 (i). This means that $u \in U_\varphi$, and so $\chi(y)y \in W$. Thus $\mathfrak{r}_\mathbf{h}^\times \subset K^\times W$, which implies that k is unramified over K. Since $K(J(\tau)) \subset K(c_2(\tau), c_3(\tau)) \subset k$ and $K(J(\tau))$ is the Hilbert class field over K, we obtain our assertion on k. By Theorem 14.9, $\beta(y) = \chi(y)$, and by (14.11), $\mu(y) = \chi(y)$ for $y \in \mathfrak{r}_\mathbf{h}^\times$. For $\alpha \in K$ prime to $\mathfrak{c}$ we have $\lambda(\alpha\mathfrak{r}) = \mu_\mathbf{a}(\alpha)\prod_{v|\mathfrak{c}} \mu_v(\alpha) = \alpha^{-1}|\alpha|\chi(\alpha)$. This completes our proof.

14.16. We can always find $\mathfrak{c}$, m, and χ as in §14.12, provided $K \neq \mathbf{Q}(i)$. Indeed, if $K \neq \mathbf{Q}(i)$, then we can find a prime p that splits in K but remains prime in $\mathbf{Q}(i)$. We can take $\mathfrak{c}$ to be a prime ideal in K dividing p. An example of C_τ over $\mathbf{Q}$ when $K = \mathbf{Q}(i)$ will be discussed in §14.19.

Once $\mathfrak{c}$, m, and χ are chosen as in (14.19), we put

$$\varphi_m(z) = (1/2)\sum_{t=1}^{m-1} \chi(t)F_1(z;\, t/m,\, 0), \tag{14.21}$$

and we can take φ_m as φ. If χ is primitive, this is the same as $\varphi_{1,\chi}^{\ell}(z)$ of (9.10a) with $N = m$, and belongs to $\mathscr{M}_1(\mathbf{Q})$.

Thus, for a given fractional ideal $\mathfrak{x} = \mathbf{Z}\alpha + \mathbf{Z}\beta$ in K and $\tau = \alpha/\beta \in H$ we can find a curve C_τ isomorphic to $\mathbf{C}/\mathfrak{x}$ and defined over the Hilbert class field of K, provided $\varphi(\tau) \neq 0$ for our choice of φ and τ as in Theorem 14.9; moreover, we know the Hecke character determined by the curve.

If we take $\mathfrak{a} = \mathbf{Z}\sigma + \mathbf{Z}r$ as in (14.20), we have a formula (13.29a), which explains the meaning of $\varphi(\sigma/r)$. If in particular K has class number 1 and $\mathfrak{r}^\times = \{\pm 1\}$, then the formula gives

$$L(1/2,\, \lambda) = 2\pi m^{-1/2}\varphi_m(\sigma) \tag{14.22}$$

with $\sigma \in \mathfrak{c} \cap H$ such that $\mathfrak{r} = \mathbf{Z}\sigma + \mathbf{Z}$.

14.17. Take for example $\tau = (3 + \sqrt{-7})/2$, $\sigma = -\tau^\rho$, and $\mathfrak{c} = \mathfrak{r}\sigma$ as in §§13.12 and 13.13. We have then $m = 4$ and φ_4 of (14.21) coincides with $F_1(z; 1/4,\, 0)$. Therefore C_σ in this case is the curve $C(\sigma\omega,\, \omega)$ of (13.34a). Put $\gamma = \sqrt{-7}\,\sigma$ and denote the curve $C(\tau\omega_2,\, \omega_2)$ of (13.34) by C. Then the map $(X,\, Y) \to f(X,\, Y) = (\gamma X,\, \gamma^{3/2}Y)$ gives an isomorphism of C onto C_σ defined over the field $K(\gamma^{1/2}) = K\bigl((-7)^{1/4}\bigr)$, as $4\sigma = (1 + \sqrt{-7})^2$.

Let μ and λ be defined as in Theorem 14.15 for the present C_σ. Let κ be the Hecke character of $K_{\mathbf{A}}^\times$ corresponding to $K(\gamma^{1/2})$. Then $\kappa(y) = (\gamma^{1/2})^{\{y\}}/\gamma^{1/2}$ for $y \in K_{\mathbf{A}}^\times$. Now, if $\xi : \mathbf{C}/\mathfrak{r} \to C_\sigma$ resp. $\xi' : \mathbf{C}/\mathfrak{r} \to C$ is the Weierstrass $\wp$-parametrization of C_σ resp. C and $\xi'(w) = (X,\, Y)$ for $w \in K$, then $\xi(w) = f(\xi'(w)) = (\gamma X,\, \gamma^{3/2}Y)$. Therefore $\xi(w)^{\{y\}} = \kappa(y)f(\xi'(w)^{\{y\}})$ for $y \in K_{\mathbf{A}}^\times$. Thus $\kappa\mu$ gives the Hecke character of $K_{\mathbf{A}}^\times$ determined by C. By Lemma 14.5 (ii), $L(s,\, \kappa^*\lambda)$ is the zeta function of C over $\mathbf{Q}$. It is not difficult to show that $\kappa^*\lambda$ has conductor $\sqrt{-7}\mathfrak{r}$, but we leave the details of the proof to the reader.

14.18. Let us now take the setting of Theorem 14.15 as follows: $K = \mathbf{Q}(\sqrt{-m}\,)$ with a squarefree positive integer m such that $0 < m-3 \in 4\mathbf{Z}$, $\chi(a) = \left(\dfrac{-m}{a}\right)$, and $\mathfrak{c} = \sqrt{-m}\,\mathfrak{r}$, we put $\sigma_m = (m + \sqrt{-m}\,)/2$ and take this as σ of the theorem. There is a character λ_0 of $(\mathfrak{r}/\mathfrak{c})^\times$ such that $\lambda_0(a) = \chi(a)$ for $a \in \mathbf{Z}$. Then we have a Hecke ideal character λ such that $\lambda(\alpha\mathfrak{r}) = \lambda_0(\alpha)\alpha^{-1}|\alpha|$ for $\alpha \in \mathfrak{r}$ prime to $\mathfrak{c}$. Define φ_m by (14.21), and assuming that $\varphi_m(\sigma_m) \neq 0$, define C_{σ_m} with φ_m as φ. By Theorem 14.15 this λ is exactly the character determined by C_{σ_m} in that theorem, and (14.22) gives $L(1/2, \lambda) = 2\pi m^{-1/2}\varphi_m(\sigma_m)$, when K has class number 1.

Put $k_0 = \mathbf{Q}\big(c_2(\sigma_m), c_3(\sigma_m)\big)$. We note that Lemma 1.6 (v) is applicable to φ_m and $E_{2\mu}$. Thus $\overline{E_{2\mu}(\sigma_m)} = E_{2\mu}(-\overline{\sigma_m}) = E_{2\mu}(\sigma_m - m) = E_{2\mu}(\sigma_m)$; similarly $\overline{\varphi_m(\sigma_m)} = \varphi_m(\sigma_m - m) = \varphi_m(\sigma_m)$, and so $\varphi_m(\sigma_m)$, $c_2(\sigma_m)$, and $c_3(\sigma_m)$ are real. Therefore $\mathbf{Q}\big(J(\sigma_m)\big) \subset k_0 \subset \mathbf{R}$, and $K \not\subset k_0$. Since $k = K\big(J(\sigma_m)\big) = Kk_0$, we have $k_0 = \mathbf{Q}\big(J(\sigma_m)\big)$, and C_{σ_m} is rational over k_0. Let μ be the Hecke character of $K_\mathbf{A}^\times$ associated with C_{σ_m}. By Theorem 14.9 and Lemma 14.5 (ii), $L(s, \mu \circ N_{k/K})$ is the zeta function of C_{σ_m} over k_0.

Let us take $m = 7$, and compare this result with what we found in §14.17. We have $k_0 = \mathbf{Q}$. We easily see that there is a unique character of conductor $\sqrt{-7}\,\mathfrak{r}$, and so μ^* coincides with $\kappa^*\lambda$ of §14.17. Therefore C_{σ_7} and C of §14.17 are both $\mathbf{Q}$-rational and have the same zeta function over $\mathbf{Q}$. Thus they are isomorphic over K by Lemma 14.5 (i). We can even show that they are isomorphic over $\mathbf{Q}$ as follows. In general the isomorphism class of a $\mathbf{Q}$-rational curve of the form $C(\tau\omega_2, \omega_2)$ over $\mathbf{Q}$ is determined by the set of numbers $\mathbf{Q}^\times\mathfrak{r}^\times\omega_2$. If $K = \mathbf{Q}(\sqrt{-7}\,)$, the set is $\mathbf{Q}^\times\omega_2$. Moreover, another such curve $C(\tau\omega_2', \omega_2')$ isomorphic to $C(\tau\omega_2, \omega_2)$ over K is not isomorphic to $C(\tau\omega_2, \omega_2)$ over $\mathbf{Q}$ exactly when $\mathbf{Q}^\times\omega_2' = \sqrt{-7}\,\mathbf{Q}^\times\omega_2$. This is a special case of Theorem 15.10. In the present setting, the set $\mathbf{Q}^\times\omega_2$ for C_{σ_7} is $\mathbf{Q}^\times 2\pi i\varphi_7(\sigma_7)$, and the set for C is $\mathbf{Q}^\times\omega_2$ with $\omega_2 = 2\pi i|\varepsilon_2(\tau)|^{1/2}$ as in (13.34). Since both $\varphi_7(\sigma_7)$ and $|\varepsilon_2(\tau)|^{1/2}$ are real, we cannot have $\sqrt{-7}$ as a quotient of the periods. Therefore C_{σ_7} must be isomorphic to the curve C of (13.34) over $\mathbf{Q}$.

14.19. Let $K = \mathbf{Q}(i)$ and $\Gamma = S \cap G$ with

$$S = \big\{x \in U \,\big|\, \det(x) - 1 \prec 4\mathbf{Z},\ \ (1, 1)x - (1, 1) \prec 4\mathbf{Z}\big\}. \tag{14.23}$$

We easily see that Γ coincides with Γ_3 of §10.7. Define q by (14.5) with $\tau = i$ and put $\mathfrak{c} = (1+i)^3\mathfrak{r}$. If $s \in \mathfrak{r}_\mathbf{h}^\times$ and $s-1 \prec \mathfrak{c}$, then $ss^\rho - 1 \prec 4\mathbf{Z}$ and $(1+i)s-(1+i) \prec 4\mathfrak{r}$, and so we see that $q(s) \in S$. We now consider the map ξ of (14.13) with $\varphi(z) = F_1(z;\, 1/4, 1/4)$. Then $\varphi \in \mathscr{M}_1(\Gamma, \mathbf{Q}(i))$ and $\varphi(i)$ becomes the basic constant $F_1(i;\, 1/4, 1/4)$ discussed in §§13.7 and 13.8. Moreover the equation of the curve C_i is $Y^2 = 4X^3 + 16X$, as $E_4(i)/F_1(i;\, 1/4, 1/4) = -4/5$; see (13.16). Given $y \in \mathfrak{r}_\mathbf{h}^\times$, there exists $\zeta \in \{\pm 1, \pm i\}$ such that $\zeta^{-1}y - 1 \prec \mathfrak{c}$. Put $s = \zeta^{-1}y$ and $r = q(s)^{-1}$. Then $r \in S$, and for $w = ai + b$ we have $s^{-1}(ai+b) = a'i + b'$, where $(a', b') = (a, b)r$. By Lemma 14.3 (ii), $\varphi^{[r]} = \varphi^{\{\det(r)^{-1}\}} = \varphi$. Therefore by (14.3), (14.6) and (14.13) we have

$$\begin{aligned}\xi(w)^{\{y\}} = \xi(w)^{\{s\}} &= \big(\varphi(i)^{-\nu}F_\nu(i;\, a, b)\big)_{\nu=2,3}^{\{s\}} \\ &= \big(\varphi(i)^{-\nu}F_\nu(i;\, a', b')\big)_{\nu=2,3} = \xi(s^{-1}w) = \xi(\zeta y^{-1}w).\end{aligned}$$

This shows that the Hecke character ψ of $K_\mathbf{A}^\times$ determined by C_i is given by $\psi(y) = \zeta$, which means that the ideal character λ of §13.7 corresponds to $\overline{\psi}$.

Take the curve $C(i\omega, \omega)$ of (13.19) with ω as in (13.20). Then the map $f : (X, Y) \mapsto (\gamma^2 X, \gamma^3 Y)$ with $\gamma = 1 + i$ gives an isomorphism of $C(i\omega, \omega)$ onto C_i rational over K, and these curves determine the same Hecke character ψ. Thus, by Lemma 14.5 (ii), $L(s, \psi)$ is their zeta function over $\mathbf{Q}$, but clearly they are not isomorphic over $\mathbf{Q}$. However, the map $x \mapsto f((1+i)x)$ gives a $\mathbf{Q}$-rational isogeny of C to C_i.

It is natural to compare $L(1/2, \lambda)$ with the "period" $2\pi i F_1(i;\, 1/4,\, 1/4)$. In fact, from (13.22) we obtain

$$L(1/2, \lambda) = -2^{-1}(1+i)\pi i F_1(i;\, 1/4,\, 1/4) = \omega/4. \tag{14.24}$$

We will explain the conceptual meaning of the factor $1 + i$ in Theorem 15.10 and §15.11.

14.20. Let us insert here a remark about the L-function of the character $\mu \circ N_{k/K}$ of Theorem 14.7. As noted in §14.6, there are exactly $[k : K]$ such characters μ of $K_{\mathbf{A}}^{\times}$ that produce the same character $\mu \circ N_{k/K}$. If we fix one such character μ, then every other character of that type is of the form $\kappa\mu$ with a character κ of $K_{\mathbf{A}}^{\times}/K^{\times}N_{k/K}(k_{\mathbf{A}}^{\times})$, and vice versa. Then we easily see that

$$L(s, \mu \circ N_{k/K}) = \prod_{\kappa} L(s, \kappa\mu), \tag{14.25}$$

where the product is over all the characters κ of $K_{\mathbf{A}}^{\times}/K^{\times}N_{k/K}(k_{\mathbf{A}}^{\times})$. Consequently $L(1/2, \mu \circ N_{k/K})$ is the product of the quantities $L(1/2, \lambda)$ with the characters λ we discussed in Section 13.

14.21. Let us next consider the case $K = \mathbf{Q}(\sigma)$ and $\mathfrak{c} = 2\sigma\mathfrak{r}$, $\sigma = \sqrt{-2}$ as in §13.16. This case is different from the cases we discussed with $K = \mathbf{Q}(\zeta)$, $\mathbf{Q}(i)$, and $\mathbf{Q}(\sqrt{-7})$, as a nontrivial abelian extension k of K appears. Put

$$\chi(a) = \left(\frac{-1}{a}\right), \qquad \varphi(z) = F_1(z;\, 1/4,\, 0). \tag{14.26}$$

Then $\varphi \in \mathscr{M}_1(\mathbf{Q})$ and $\varphi\|_1\gamma = \chi(a_\gamma)\varphi$ for every $\gamma \in \Gamma^0(4)$. Therefore, by Lemma 14.3 (i), $\varphi^{[x]} = \chi(a_x)\varphi$ for every $x \in U^0(4)$. Define q by (14.5) with σ as τ; define also W and k as in Theorem 14.9. We have $q(b\sigma + a) = \begin{bmatrix} a & -2b \\ b & a \end{bmatrix}$. Clearly $\mathfrak{r}_p^{\times} \subset W$ for every prime $p \neq 2$. At the prime 2, put $\mathfrak{c}_2 = 2\sigma\mathfrak{r}_2$. Then we easily find that if $x \in q(1 + \mathfrak{c}_2)$, then $x \in U^0(4)$ and $\chi(a_x) = 1$, and so $1 + \mathfrak{c}_2 \subset W$. Let ε be the projection of $1 + \sigma$ to K_2. Since $-\varepsilon^2 \in 1 + \mathfrak{c}_2$, we see that $\varepsilon^2 \in K^{\times}W$ and $\beta(\varepsilon^2) = -1$, where β is as in Theorem 14.9. If $\varepsilon \in K^{\times}W$, then $\beta(\varepsilon)^2 = -1$, which is a contradiction, as the values of β must belong to K. Thus $\varepsilon \notin K^{\times}W$. Now $\mathfrak{r}_2^{\times}$ is generated by ε and $1 + \mathfrak{c}_2$. Since $K_{\mathbf{A}}^{\times} = K^{\times}K_{\mathbf{a}}^{\times}\mathfrak{r}_{\mathbf{h}}^{\times}$ and $\varepsilon^2 \in W$, we have $[K_{\mathbf{A}}^{\times} : K^{\times}W] = 2$, and so $[k : K] = 2$. Now from (13.45b, e, f) we see that $\mathbf{Q}(c_2(\sigma), c_3(\sigma)) = \mathbf{Q}(\sqrt{2})$, and so $k = K(\sqrt{2}) = K(i)$.

Define C_σ with φ of (14.26) and μ by (14.11). Then μ corresponds to the character λ of §13.17. (Recall that μ is defined only on $K^\times W$ and there are two extensions of μ to $K_{\mathbf{A}}^\times$.) By Theorem 14.9 and Lemma 14.5 (ii), the zeta function of C_σ over $\mathbf{Q}(\sqrt{2})$ is $L(s, \lambda \circ N_{k/K})$. By (14.25) we have $L(s, \lambda \circ N_{k/K}) = L(s, \lambda)L(s, \kappa\lambda)$, where κ is the quadratic Hecke character of K corresponding to the extension k. We easily see that $\kappa(\alpha\mathfrak{r}) = \chi(\alpha\alpha^\rho)$ for $\alpha \in \mathfrak{r}$ prime to $\mathfrak{c}$ and $\kappa\lambda = \overline{\lambda}$. We gave $L(1/2, \lambda)$ in (13.46), and we can similarly determine $L(1/2, \overline{\lambda})$ by replacing i by $-i$ in (13.42).

So far we have considered only the members of the family $\{C_z\}$ of (14.12). In a more general case we can state the following result:

Theorem 14.22. *Let C be an elliptic curve which is defined over $\overline{\mathbf{Q}}$ and isomorphic to $\mathbf{C}/(\mathbf{Z}\tau + \mathbf{Z})$ with $\tau \in K \cap H$, where K is as in §14.4. Let η be a nonzero holomorphic 1-form on C defined over $\overline{\mathbf{Q}}$, and φ an element of $\mathscr{A}_1(\overline{\mathbf{Q}})$ such that $\varphi(\tau) \neq 0$. Then $\int_c \eta \in \overline{\mathbf{Q}}\pi\varphi(\tau)$ for every $c \in H_1(C, \mathbf{Z})$.*

Proof. Define C_z by (14.12) with the present φ. Let $\eta' = dX/Y$ defined on C_τ. Since C_τ is K_{ab}-rational and isomorphic to $\mathbf{C}/(\mathbf{Z}\tau + \mathbf{Z})$, we can find a $\overline{\mathbf{Q}}$-rational isomorphism f of C_τ onto C. Then $\eta \circ f = a\eta'$ with $a \in \overline{\mathbf{Q}}^\times$. Therefore our problem can be reduced to C_τ. Since $C_\tau = C(\tau\omega, \omega)$ with $\omega = 2\pi i\varphi(\tau)$, for every $c \in H_1(C_\tau, \mathbf{Z})$ we have $\int_c \eta' \in \mathbf{Z}\tau\omega + \mathbf{Z}\omega \subset K\omega$, from which we obtain the desired fact.

The above theorem can be generalized to the case of abelian varieties with complex multiplication; see [S98, Sections 30, 31, and 32].

14.23. Put $\eta(z) = \Delta(z)^{1/24} = \mathbf{q}^{1/24}\prod_{n=1}^{\infty}(1 - \mathbf{q}^n)$, $\mathbf{q} = \mathbf{e}(z)$, as usual. As we already mentioned in §14.12, we can take this η as the function φ. To study this case, put

$$\gamma_2(z) = \frac{12g_2(z)}{(2\pi)^4\eta(z)^8}, \qquad \gamma_3(z) = \frac{\sqrt{27}g_3(z)}{(2\pi)^6\eta(z)^{12}}. \tag{14.27}$$

Then the curve C_z of (14.12) takes the form

$$Y^2 = 4X^3 - \gamma_2(z)X - 8\gamma_3(z). \tag{14.28}$$

We can easily verify that

$$\gamma_2(z)^3 = j(z), \qquad \gamma_3(z)^2 = j(z) - 12^3, \tag{14.29a}$$

$$\gamma_2(z+1) = \mathbf{e}(-1/3)\gamma_2(z), \quad \gamma_2(-z^{-1}) = \gamma_2(z), \tag{14.29b}$$

$$\gamma_3(z+1) = -\gamma_3(z), \quad \gamma_3(-z^{-1}) = -\gamma_3(z). \tag{14.29c}$$

The functions γ_2 and γ_3 were introduced by Weber, and the numerical values of $\gamma_\nu(\tau)$ for many imaginary quadratic τ were given in [Web]. We already employed such values in §§13.13 and 13.17. Several explicit examples of curves of the type (14.28) and the Hecke characters associated with them were given by Rumely in [Ru]. It may be an interesting exercise to find more examples of curves and L-functions by using such results of Weber and others, and to compare $L(1/2, \lambda)$ with the periods.

14.24. The curve C_τ for $\tau \in K \cap H$ with any choice of $\varphi \in \mathscr{A}_1(\mathbf{Q}_{\mathrm{ab}})$ has the property that its torsion points are rational over K_{ab}. Therefore, if $K = \mathbf{Q}(\sqrt{-\ell})$ with a prime $\ell \equiv 1 \pmod 8$ and $\mathbf{Z}\tau + \mathbf{Z}$ is a fractional ideal, then for any choice of φ the field $\mathbf{Q}(c_2(\tau), c_3(\tau))$ cannot be the field of moduli $\mathbf{Q}(j(\tau))$; see the last statement of Section 5 of [S71b].

14.25. Given λ as in §13.5, put

$$f(z) = \sum_{\mathfrak{x}} \lambda(\mathfrak{x}) N(\mathfrak{x})^{1/2} \mathbf{e}(N(\mathfrak{x})z) \qquad (z \in H), \tag{14.30}$$

where $\mathfrak{x}$ runs over the integral ideals in K. Then we can show that f is a cusp form of weight $\nu + 1$, and

$$\int_0^\infty f(iy) y^{s-1} dy = (2\pi)^{-s} \Gamma(s) L(s - \nu/2, \lambda). \tag{14.31}$$

Thus we can study the value $L(k/2, \lambda)$ as special cases of the values of the Mellin transform of a cusp form. For a full discussion of this topic, the reader is referred to [S76b], [S77], and [S78c]. In this book, however, we have restricted ourselves to the case of Hecke characters of an imaginary quadratic field and concentrated on the problem of how to find numerical values. We also view the periods as the values of modular forms of weight 1, as we think that such is more natural and practical than the periods of the cusp form of (14.30) in the case of $L(s, \lambda)$ of our type, and also that this aspect of the periods has been neglected by previous researchers.

CHAPTER VI

SUPPLEMENTARY RESULTS

15. Isomorphism classes of abelian varieties with complex multiplication

15A. The general case

15.1. The isomorphism class of an elliptic curve with complex multiplication depends on the choice of a field of definition over which isomorphisms are considered. In the first part of this section we will investigate some problems concerning such isomorphism classes more generally for abelian varieties with complex multiplication. The case of elliptic curves will be treated in the second part. Hereafter all algebraic number fields in this section are subfields of $\mathbf{C}$. We let ρ denote complex conjugation and its restriction to any subfield of $\mathbf{C}$.

Let F be a totally real algebraic number field and K a totally imaginary quadratic extension of F. We call such a K a **CM-field.** By a **CM-type** we mean a pair $(K, \varPhi)$ consisting of K and a set $\varPhi = \{\varphi_\nu\}_{\nu=1}^n$, where $n = [K : \mathbf{Q}]/2$ and the φ_ν are ring-injections of K into $\mathbf{C}$ such that $\{\varphi_\nu, \varphi_\nu\rho\}_{\nu=1}^n$ is the set of all ring-injections of K into $\mathbf{C}$. Once $(K, \varPhi)$ is fixed, we denote by K^* the field generated over $\mathbf{Q}$ by $\sum_{\nu=1}^n \alpha^{\varphi_\nu}$ for all $\alpha \in K$. This of course depends on $\varPhi$, but there will be no confusion, as we consider K^* with a fixed $(K, \varPhi)$. It is known that K^* is a CM-field. For these and other basic properties of CM-fields and CM-types, the reader is referred to [S98, §§18.1 and 18.2]. In particular, if φ is a ring-injection of a CM-field into $\mathbf{C}$, then $\varphi\rho = \rho\varphi$.

We now consider a structure (A, ι) consisting of an abelian variety A defined over a subfield of $\mathbf{C}$ and a ring-injection ι of K into $\mathrm{End}(A) \otimes \mathbf{Q}$. We view A as the set of points with coordinates in $\mathbf{C}$, but say that A is defined (or rational) over a subfield k of $\mathbf{C}$ if A is a projective variety defined over k. Let α be an element of K such that $\iota(\alpha) \in \mathrm{End}(A)$. Then $\iota(\alpha)$ acts on the space of holomorphic 1-forms on A. This action can be extended to a representation of K on that space. We say that (A, ι) is of type $(K, \varPhi)$ if that representation is equivalent over $\mathbf{C}$ to $\alpha \mapsto \mathrm{diag}[\alpha^{\varphi_1}, \ldots, \alpha^{\varphi_n}]$.

Let $\mathfrak{o}$ be an order in K that is not necessarily maximal. Put

$$W(\mathfrak{o}) = \{x \in \mathfrak{o}^\times \mid xx^\rho = 1\}, \qquad W_0(\mathfrak{o}) = \{x/x^\rho \mid x \in \mathfrak{o}^\times\}.$$

Since K is a CM-field, for $x \in W(\mathfrak{o})$ and every ring-injection φ of K into $\mathbf{C}$ we have $|x^\varphi|^2 = x^\varphi x^{\varphi\rho} = x^\varphi x^{\rho\varphi} = (xx^\rho)^\varphi = 1$. Such an algebraic integer x must be a root of unity. Thus $W(\mathfrak{o})$ consists of the roots of unity contained in $\mathfrak{o}$. Clearly $W_0(\mathfrak{o}) \subset W(\mathfrak{o})$.

Lemma 15.2. *We have* $[W(\mathfrak{o}) : W_0(\mathfrak{o})][\mathfrak{o}^\times : (\mathfrak{o}^\times \cap F)W(\mathfrak{o})] = 2$. *Consequently* $[W(\mathfrak{o}) : W_0(\mathfrak{o})] \leq 2$, *and* $[W(\mathfrak{o}) : W_0(\mathfrak{o})] = 2$ *if and only if* $\mathfrak{o}^\times = (\mathfrak{o}^\times \cap F)W(\mathfrak{o})$.

PROOF. For simplicity we will write W and W_0 for $W(\mathfrak{o})$ and $W_0(\mathfrak{o})$. Put also $W^2 = \{\zeta^2 \mid \zeta \in W\}$. Since $\zeta^2 = \zeta/\zeta^\rho$ for $\zeta \in W$, we have $W^2 \subset W_0 \subset W$, and $[W : W^2] = 2$, as W is cyclic and contains ± 1. Put $g(x) = x/x^\rho$ for $x \in \mathfrak{o}^\times$. Then $g(\mathfrak{o}^\times) = W_0$ and $g((\mathfrak{o}^\times \cap F)W) = W^2$. Suppose $g(x) \in W^2$ with $x \in \mathfrak{o}^\times$. Then $x/x^\rho = \zeta^2$ with $\zeta \in W$, and so $(\zeta^{-1}x)^\rho = \zeta^{-1}x$. Thus $\zeta^{-1}x \in F$, which shows that $x \in (\mathfrak{o}^\times \cap F)W$. Therefore $[W_0 : W^2] = [\mathfrak{o}^\times : (\mathfrak{o}^\times \cap F)W]$. This proves our lemma.

15.3. If $F = \mathbf{Q}$, then K is imaginary quadratic and $\mathfrak{o}^\times = W(\mathfrak{o})$, and so $[W(\mathfrak{o}) : W_0(\mathfrak{o})] = 2$. However, if $F \neq \mathbf{Q}$, it can happen that $W(\mathfrak{o}) = W_0(\mathfrak{o})$. For instance, take a real quadratic field F with a totally positive fundamental unit ε. Let $K = F(\eta)$ with $\eta^2 = -\varepsilon$ and let $\mathfrak{o}$ be an order in K containing η. Then $-1 = \eta/\eta^\rho \in W_0$. If $W \neq W_0$, then $W_0 = W^2$, and so $-1 = i^2$ with an element i of W. Thus $i\eta \in F$, as $(i\eta)^\rho = i\eta$, and $\varepsilon = (i\eta)^2$, which is a contradiction. Therefore $W(\mathfrak{o}) = W_0(\mathfrak{o})$ in this case.

Given (A, ι) and (A', ι') with the same K, by an **isogeny** resp. **isomorphism** of (A, ι) to (A', ι'), we mean an isogeny resp. isomorphism f of A to A' such that $f\iota(\alpha) = \iota'(\alpha)f$ for every $\alpha \in K$. We say that (A, ι) is rational, or defined, over k, or simply k-rational, if A and $\iota(\alpha)$ for every $\alpha \in K$ such that $\iota(\alpha) \in \mathrm{End}(A)$ are defined over k.

A CM-type (K, Φ) is called **primitive** if $\iota(K) = \mathrm{End}(A) \otimes \mathbf{Q}$ for every (A, ι) of type (K, Φ). This is so if and only if $\iota(K) = \mathrm{End}(A) \otimes \mathbf{Q}$ for any single choice of (A, ι) of type (K, Φ). Also, for (A, ι) of type (K, Φ), we have $\iota(K) = \mathrm{End}(A) \otimes \mathbf{Q}$ if and only if A is **simple,** *that is, A has no nontrivial abelian subvariety (over* $\mathbf{C}$*) other than itself. For these, the reader is referred to [S98, Section 8].*

Given (A, ι) of type (K, Φ), we denote by ι_F the restriction of ι to F, and define the k-rationality of (A, ι_F) as a natural analogue of that of (A, ι).

Lemma 15.4. *Suppose (K, Φ) is primitive. Let (A, ι) be of type (K, Φ). Then the following assertions hold*

(i) *If (A', ι') is of type (K, Φ), then every isogeny of A to A' is an isogeny of (A, ι) to (A', ι').*

(ii) *Suppose A is defined over an algebraic number field k; then (A, ι) is k-rational if and only if $K^* \subset k$.*

(iii) *If* $(A,\ \iota_F)$ *is rational over an algebraic number field* h *not containing* K^*, *then* $[hK^*:h]=2$ *and* $h\cap K^*$ *is the maximal real subfield of* K^*.

(iv) *If* A *is defined over an algebraic number field* h *such that* $h\cap K^*$ *is the maximal real subfield of* K^*, *then* $[hK^*:h]=2$, $(A,\ \iota_F)$ *is* h*-rational, and* $\iota(\alpha)^\sigma=\iota(\alpha^\rho)$, *where* σ *is the generator of* $\mathrm{Gal}(hK^*/h)$.

PROOF. Assertion (i) is included in [S98, §14.1, Proposition 1], (ii) in [S98, §8.5, Proposition 30], and (iii) in [S98, 20.4, (iv)]. To prove (iv), put $k=hK^*$. Then clearly $[k:h]=2$ and $\sigma=\rho$ on K^*; also, $(A,\ \iota)$ is k-rational by (ii). Define $\iota_1: K\to \mathrm{End}(A)\otimes\mathbf{Q}$ by $\iota_1(\alpha)=\iota(\alpha^\rho)^\sigma$ with σ as in (iv). Now the representation of $\iota_1(\alpha)$ on the space of holomorphic 1-forms has eigenvalues $\{\alpha^{\rho\varphi_\nu\sigma}\}_{\nu=1}^n$. For every $m\in\mathbf{Z},>0$, and $\alpha\in K$ we have $\sum_{\nu=1}^n(\alpha^m)^{\varphi_\nu}\in K^*$, and so $\sum_{\nu=1}^n(\alpha^{\rho\varphi_\nu\sigma})^m=\sum_{\nu=1}^n(\alpha^{\varphi_\nu\rho\sigma})^m=\left\{\sum_{\nu=1}^n(\alpha^m)^{\varphi_\nu}\right\}^{\rho\sigma}$, which equals $\sum_{\nu=1}^n(\alpha^m)^{\varphi_\nu}=\sum_{\nu=1}^n(\alpha^{\varphi_\nu})^m$, as $\rho=\sigma$ on K^*. Consequently $\{\alpha^{\rho\varphi_\nu\sigma}\}_{\nu=1}^n=\{\alpha^{\varphi_\nu}\}_{\nu=1}^n$, which means that $(A,\ \iota_1)$ is of type $(K,\ \Phi)$. Applying (i) to the identity map of A to itself, we see that $\iota_1=\iota$. Thus $\iota(\alpha^\rho)=\iota(\alpha)^\sigma$, and so $\iota(\alpha)=\iota(\alpha)^\sigma$ if $\alpha\in F$, which means that $(A,\ \iota_F)$ is h-rational. This completes the proof.

15.5. We now fix a primitive CM-type $(K,\ \Phi)$. If $(A,\ \iota)$ is of type $(K,\ \Phi)$, then A is isomorphic as a complex manifold to $\mathbf{C}^n/p(\mathfrak{a})$ with a $\mathbf{Z}$-lattice $\mathfrak{a}$ in K, where p is the injection of K into $\mathbf{C}^n$ defined by $p(\alpha)=(\alpha^{\varphi_\nu})_{\nu=1}^n$; see [S98, §6.2]. Then we say that $(A,\ \iota)$ is of type $(K,\ \Phi,\ \mathfrak{a})$. Clearly the set of data $(K,\ \Phi,\ \mathfrak{a})$ determines the isomorphism class of $(A,\ \iota)$ over $\mathbf{C}$. The lattice $\mathfrak{a}$ can be replaced by $\beta\mathfrak{a}$ with any $\beta\in K^\times$. Also, if

$$\mathfrak{o}=\{\gamma\in K\mid \gamma\mathfrak{a}\subset\mathfrak{a}\},\tag{15.1}$$

then $\iota(\mathfrak{o})=\iota(K)\cap\mathrm{End}(A)$. Now suppose $(A,\ \iota)$ of type $(K,\ \Phi,\ \mathfrak{a})$ is rational over an algebraic number field k. Let $k_{\mathbf{A}}^\times$ denote the idele group of k. Then $(A,\ \iota)$ determines a Hecke character ψ of $k_{\mathbf{A}}^\times$ by which we can describe the zeta function of A over k. This ψ is defined by natural generalizations of (14.7a, b) and (14.8); see [S98, 19.8 and 19.10]. We do not make precise statements here, as they are unnecessary for our treatment. We denote by $\Psi(K,\ \Phi,\ \mathfrak{a},\ k)$ the set of all such characters. Then every element ψ of $\Psi(K,\ \Phi,\ \mathfrak{a},\ k)$ determines the isomorphism class of $(A,\ \iota)$ over k, and vice versa; see [S98, 19.11 and 19.12]. We have $K^*\subset k$ by Lemma 15.4 (ii), but often $(A,\ \iota_F)$ can be defined over a field not containing K^*. We therefore ask whether the isomorphism class of $(A,\ \iota_F)$ can still be determined by ψ.

To answer this question, we take the following setting. We fix an algebraic number field h satisfying the following condition:

(15.2) $h\cap K^*$ *is the maximal real subfield of* K^*.

Given such an h, we put $k=hK^*$, fix $\psi\in\Psi(K,\ \Phi,\ \mathfrak{a},\ k)$, and let $\Xi(\mathfrak{a},\ \psi,\ h)$ denote the set of all k-rational $(A,\ \iota)$ of type $(K,\ \Phi,\ \mathfrak{a},\ k)$ such that A is h-

rational and (A, ι) determines ψ. By Lemma 15.4 (iv), (A, ι_F) is h-rational for $(A, \iota) \in \Xi(\mathfrak{a}, \psi, h)$. From Lemma 15.4 (i) we see that (A, ι_F) is isomorphic to (A', ι'_F) over h if and only if A and A' are isomorphic over h. Thus, if (A, ι) is taken from $\Xi(\mathfrak{a}, \psi, h)$, then the isomorphism class of A over h is practically the same as the isomorphism class of (A, ι_F) over h. Notice also that all the members of $\Xi(\mathfrak{a}, \psi, h)$ are isomorphic over k, as mentioned above.

Theorem 15.6. *Define* $\mathfrak{o}$ *by (15.1). If* $\Xi(\mathfrak{a}, \psi, h)$ *is nonempty, the following asssertions hold:*

(i) *The number of the isomorphism classes over* h *of the structures* (A, ι_F) *obtained from* $(A, \iota) \in \Xi(\mathfrak{a}, \psi, h)$ *is exactly* $[W(\mathfrak{o}) : W_0(\mathfrak{o})]$.

(ii) *All such structures* (A, ι_F) *are isogenous over* h.

PROOF. For simplicity we will write W and W_0 for $W(\mathfrak{o})$ and $W_0(\mathfrak{o})$. We prove (i) in four steps.

(1) Let σ be the generator of $\mathrm{Gal}(k/h)$. Let us fix a member (A, ι) of $\Xi(\mathfrak{a}, \psi, h)$ and let $(A', \iota') \in \Xi(\mathfrak{a}, \psi, h)$. Then $\iota(\alpha)^\sigma = \iota(\alpha^\rho)$ and $\iota'(\alpha)^\sigma = \iota'(\alpha^\rho)$ for every $\alpha \in \mathfrak{o}$ by Lemma 15.4 (iv). Take a k-rational isomorphism f of (A, ι) onto (A', ι'). Then we easily see that f^σ is also such an isomorphism, and so $f^\sigma = f \circ \iota(\varepsilon)$ with $\varepsilon \in \mathfrak{o}^\times$. We have $f = (f^\sigma)^\sigma = f^\sigma \circ \iota(\varepsilon)^\sigma = f \circ \iota(\varepsilon\varepsilon^\rho)$, and so $\varepsilon\varepsilon^\rho = 1$; thus $\varepsilon \in W$. Suppose $\varepsilon \in W_0$. Then $\varepsilon = \eta/\eta^\rho$ with $\eta \in \mathfrak{o}^\times$. Put $g = f \circ \iota(\eta)$. Then g is k-rational and $g^\sigma = f^\sigma \circ \iota(\eta^\rho) = f \circ \iota(\varepsilon\eta^\rho) = g$. Thus g gives an h-rational isomorphism of (A, ι_F) onto (A', ι'_F).

(2) Next suppose that $f^\sigma = f \circ \iota(\varepsilon)$ with $\varepsilon \notin W_0$. Suppose that there exists an h-rational isomorphism p of (A, ι_F) onto (A', ι'_F). Then by Lemma 15.4 (i), p gives an isomorphism of (A, ι) onto (A', ι'), and so $p = f \circ \iota(\gamma)$ with $\gamma \in \mathfrak{o}^\times$. Thus $p = p^\sigma = f^\sigma \circ \iota(\gamma^\rho) = f \circ \iota(\varepsilon\gamma^\rho)$, and so $\varepsilon = \gamma/\gamma^\rho \in W_0$, a contradiction.

(3) Given another member (A'', ι'') of $\Xi(\mathfrak{a}, \psi, h)$, take a k-rational isomorphism f_1 of (A, ι) to (A'', ι'') and put $f_1^\sigma = f_1 \circ \iota(\varepsilon_1)$ with $\varepsilon_1 \in \mathfrak{o}^\times$. Put also $g = f_1 \circ f^{-1}$. Then g is a k-rational isomorphism of (A', ι') to (A'', ι'') and $g^\sigma = g \circ \iota'(\varepsilon_1/\varepsilon)$. Applying the results of (1) and (2) to g, we see that (A', ι'_F) is isomorphic to (A'', ι''_F) over h if and only if $\varepsilon_1/\varepsilon \in W$.

(4) Suppose $[W : W_0] = 2$; take $\zeta \in W, \notin W_0$. Put $\varphi = \iota(\zeta)$. Then φ is a k-rational isomorphism of (A, ι_F) onto $(A, \iota_F) = (A, \iota_F)^\sigma$, and $\varphi^\sigma \circ \varphi = \iota(\zeta^\rho\zeta) = \iota(1)$. Therefore the simplest case of Weil's descent criterion guarantees a structure (A', ι') such that (A', ι'_F) is h-rational and a k-rational isomorphism f of (A, ι) to (A', ι') such that $f = f^\sigma \circ \varphi$. Then $f^\sigma = f \circ \iota(\zeta^{-1})$. Combining this with (3), we obtain (i).

To prove (ii), take (A', ι'), f, and ε as in (1). Since $\varepsilon\varepsilon^\rho = 1$, we can find an element α of $K^\times$ such that $\varepsilon = \alpha/\alpha^\rho$. Replacing α by its suitable multiple, we may assume that $\alpha \in \mathfrak{o}$. Put $g = f \circ \iota(\alpha)$. Then $g^\sigma = f^\sigma \circ \iota(\alpha^\rho) = f \circ \iota(\varepsilon\alpha^\rho) = g$. Thus g is an h-rational isogeny of (A, ι) to (A', ι'). This completes the proof.

Lemma 15.7. *Given (A, ι) as in §15.5, let C be a polarization of A. Then the following assertions hold:*

(i) *The field of moduli of (A, C, ι) does not depend on the choice of C.*

(ii) *The field of moduli of (A, C, ι_F) does not depend on the choice of C.*

(iii) *If K has no nontrivial automorphism other than ρ, then the field of moduli of (A, C) coincides with the field of moduli of (A, C, ι_F).*

(iv) *If (A, ι_F) is rational over a field k, then C contains a divisor rational over k.*

As to the definition of a polarization and the field of moduli, the reader is referred to [S98, Sections 4 and 17, Proposition 17.2, in particular].

PROOF. The last assertion is [S98, 20.17]. To prove (ii), take another polarization C_1 of A. Let X resp. Y be a basic polar divisor in C resp. C_1; let φ and φ_1 be the corresponding isogenies of A into the Picard variety of A; see [S98, §1.3]. Then $\varphi_1 = \varphi \circ \iota(a)$ with $a \in F$ as shown in [S98, §14.2, Proposition 2]. Let M resp. M' be the field of moduli of (A, C, ι_F) resp. (A, C_1, ι_F). Take $\sigma \in \mathrm{Aut}(\mathbf{C}/M)$. Then there exists an isomorphism λ of (A, C, ι_F) to $(A, C, \iota_F)^\sigma$ and $\lambda(X)$ is algebraically equivalent to X^σ, and so $\varphi = {}^t\lambda \circ \varphi^\sigma \circ \lambda$ by [S98, p. 6, (7)]. Thus $\varphi_1 = {}^t\lambda \circ \varphi^\sigma \circ \lambda \circ \iota(a) = {}^t\lambda \circ \varphi^\sigma \circ \iota(a)^\sigma \circ \lambda = {}^t\lambda \circ \varphi_1^\sigma \circ \lambda$, which means that $\lambda(Y)$ is algebraically equivalent to Y^σ. Thus λ is an isomorphism of (A, C_1, ι_F) to $(A, C_1, \iota_F)^\sigma$, and so σ is the identity map on M'. This shows that $M' \subset M$. Similarly, $M \subset M'$, and so $M = M'$, which is (ii). We can similarly prove (i). In fact, MK^* is the field of moduli of (A, C, ι) as shown in [S98, 20.3, (ii)]. (Take F to be D there; then K is E there.) To prove (iii), suppose there is an isomorphism f of (A, C) to $(A, C)^\sigma$ with $\sigma \in \mathrm{Aut}(\mathbf{C})$. Then there exists an automorphism γ of K such that $f \circ \iota(\alpha) = \iota(\alpha^\gamma)^\sigma \circ f$ for every $\alpha \in K$. If K has no nontrivial automorphism other than ρ, then $f \circ \iota(\alpha) = \iota(\alpha)^\sigma \circ f$ for every $\alpha \in F$, and so f is an isomorphism of (A, C, ι_F) to $(A, C, \iota_F)^\sigma$. Thus σ is the identity map on the field of moduli of (A, C, ι_F). From this we obtain (iii).

15.8. Let (A, C, ι) be as in Lemma 15.7 and let $k = hK^*$ with h satisfying (15.2). In [S71b, Theorem 9] we showed that there exist exactly two isomorphism classes of (A, C, ι_F) over h for a fixed isomorphism class of (A, C, ι) over k. However, as Theorem 15.6 shows, it can happen that the structures (A, ι_F) for such two isomorphism classes of (A, C, ι_F) form one isomorphism class over h. The difference is caused by the fact that $\iota(\varepsilon)$ for $\varepsilon \in \mathfrak{o}^\times$ is an automorphism of (A, ι), but not an automorphism of (A, C, ι) unless $\varepsilon\varepsilon^\rho = 1$.

15B. The case of elliptic curves

15.9. In the one-dimensional case, K is an imaginary quadratic field embedded in $\mathbf{C}$. We take the identity embedding of K into $\mathbf{C}$ as Φ, and so Φ can be disregarded and $K^* = K$. Also, an elliptic curve has a unique polarization, which can naturally be disregarded. Thus an elliptic curve of type $(K, \mathfrak{a})$ for a $\mathbf{Z}$-lattice $\mathfrak{a}$ is isomorphic to $\mathbf{C}/\mathfrak{a}$. We fix an algebraic number field h containing the j-invariant of $\mathbf{C}/\mathfrak{a}$ but not K, and put $k = hK$; we define $\mathfrak{o}$ by (15.1). The isomorphism class of E over k determines a Hecke character ψ of $k_{\mathbf{A}}^{\times}$, and vice versa.

We now take an h-rational elliptic curve E isomorphic to $\mathbf{C}/\mathfrak{a}$, and denote by $\iota(a)$ for $a \in K$ the element of $\mathrm{End}(E) \otimes \mathbf{Q}$ corresponding to multiplication by a on $\mathbf{C}$. We can let $\iota(K)$ act on the homology group $H_1(E, \mathbf{Q})$. Then there is an element c of $H_1(E, \mathbf{Q})$ such that $H_1(E, \mathbf{Z}) = \iota(\mathfrak{a})c$. We call such a c a **basic cycle** on E relative to $\mathfrak{a}$. (Strictly speaking, it is a homology class.) Once $\mathfrak{a}$ is fixed, $\iota(\mathfrak{o}^{\times})c$ is determined. (If we identify $H_1(E, \mathbf{Z})$ and $H_1(E, \mathbf{Q})$ with $\mathfrak{a}$ and K, then c is represented by 1.) Take an h-rational holomorphic 1-form η on E. We then put

$$\Omega_{\mathfrak{a}}(E) = h^{\times}\mathfrak{o}^{\times}\omega_0, \quad \omega_0 = \int_c \eta, \tag{15.3}$$

and call $\Omega_{\mathfrak{a}}(E)$ the set of **basic periods** on E relative to $\mathfrak{a}$. Clearly this set is independent of the choice of c and η, and it is determined by the isomorphism class of E over h, but it depends on $\mathfrak{a}$. Indeed, if we replace $\mathfrak{a}$ by $\beta\mathfrak{a}$ with $\beta \in K^{\times}$, then c must be replaced by $\iota(\beta^{-1})c$. Thus

$$\Omega_{\beta\mathfrak{a}}(E) = \beta^{-1}\Omega_{\mathfrak{a}}(E) \qquad (\beta \in K^{\times}). \tag{15.4}$$

However, our idea described in the following theorem is to compare $\Omega_{\mathfrak{a}}(E)$ with $\Omega_{\mathfrak{a}}(E')$ for E' isomorphic to E with a fixed $\mathfrak{a}$. As can easily be seen, once $\mathfrak{a}$ is chosen, the stated result is meaningful, and it does not depend on the choice of $\mathfrak{a}$.

Theorem 15.10. *With $\mathfrak{a}$, h, k, and ψ as above, let $\Xi(\mathfrak{a}, \psi, h)$ be the set of all elliptic curves E that are h-rational, isomorphic to $\mathbf{C}/\mathfrak{a}$, and determine ψ over k. Then the following assertions hold:*

(i) *All members of $\Xi(\mathfrak{a}, \psi, h)$ are isogenous over h, and divided exactly into two isomorhism classes over h.*

(ii) *Let e be the order of $\mathfrak{o}^{\times}$ and let $Y = \{\gamma \in k^{\times} \mid \gamma^e \in h\}$. Then $[Y : h^{\times}\mathfrak{o}^{\times}] = 2$.*

(iii) *If $E, E' \in \Xi(\mathfrak{a}, \psi, h)$, then $\gamma\Omega_{\mathfrak{a}}(E') = \Omega_{\mathfrak{a}}(E)$ with some $\gamma \in Y$. Moreover,*

$$E' \text{ is isomorphic to } E \text{ over } h \iff \Omega_{\mathfrak{a}}(E') = \Omega_{\mathfrak{a}}(E) \iff \gamma \in h^{\times}\mathfrak{o}^{\times}.$$

(iv) *$Y/h^{\times}\mathfrak{o}^{\times}$ is represented by*

$$\{1, 1+i\} \quad \text{if} \quad \mathfrak{o} = \mathbf{Z}[i],$$

$$\{1, \sqrt{-m}\} \quad \text{if} \quad \mathfrak{o} \neq \mathbf{Z}[i],\ K = \mathbf{Q}(\sqrt{-m}),\ 0 < m \in \mathbf{Z}.$$

PROOF. Assertion (i) is a special case of Theorem 15.6. To prove the remaining part, fix E in $\Xi(\mathfrak{a}, \psi, h)$, take any $E' \in \Xi(\mathfrak{a}, \psi, h)$, and write E and E' in the forms

$$E : y^2 = 4x^3 - g_2x - g_3; \qquad E' : y^2 = 4x^3 - g_2'x - g_3',$$

with g_ν, g_ν' in h. It is well known that every isomorphism of E' to E can be given by $\varphi_\gamma : (x, y) \mapsto (\gamma^2 x, \gamma^3 y)$ with a constant γ such that $g_2 = \gamma^4 g_2'$ and $g_3 = \gamma^6 g_3'$. We have $e = 4$ if and only if $g_3 = 0$ and $e = 6$ if and only if $g_2 = 0$; $e = 2$ if $g_2g_3 \neq 0$. Thus we see that $\gamma^e \in h$, and so $\gamma \in Y$. If $\eta = dx/y$ on E, then φ_γ sends η back to $\gamma^{-1}dx/y$ on E'. Thus $\Omega_\mathfrak{a}(E') = \gamma\Omega_\mathfrak{a}(E)$. Take another $E'' \in \Xi(\mathfrak{a}, \psi, h)$ and $\varphi_\varepsilon : E'' \to E$ with $\varepsilon \in Y$. Suppose $\Omega_\mathfrak{a}(E') = \Omega_\mathfrak{a}(E'')$. Our task is to show that E' and E'' are isomorphic over h. Since $\varepsilon\Omega_\mathfrak{a}(E) = \gamma\Omega_\mathfrak{a}(E)$, we can put $\varepsilon = \gamma b\zeta$ with $b \in h^\times$ and $\zeta \in \mathfrak{o}^\times$. Replacing φ_γ by $\varphi_{\gamma b}$ and E' by a suitable elliptic curve isomorphic to E' over h, we may assume that $b = 1$, so that $\varepsilon = \gamma\zeta$. Now $\iota(\zeta^{-1})$ sends dx/y to $\zeta^{-1}dx/y$, and so $\iota(\zeta^{-1})\circ\varphi_\gamma$ sends η back to $(\gamma\zeta)^{-1}dx/y$ on E'. Thus $\iota(\zeta^{-1})\circ\varphi_\gamma = \varphi_{\gamma\zeta}$. Since $\varepsilon = \gamma\zeta$, we see that E' and E'' are isomorphic over h. This proves (iii). Now every element γ of Y defines an isomorphism φ_γ of the above type, and so we see that $Y/h^\times\mathfrak{o}^\times$ corresponds to the isomorphism classes over h. Therefore from (i) we obtain $[Y : h^\times\mathfrak{o}^\times] = 2$, though this can be proved more directly in an elementary way. The last assertion is an easy exercise, which is left to the reader.

15.11. We add two remarks.

(A) We see from (iv) that the two isomorphism classes over h for the curves in a fixed $\Xi(\mathfrak{a}, \psi, h)$ can be represented by a pair $\{E, E'\}$ given as follows:

$$\text{(15.5a)} \qquad E : y^2 = 4x^3 - g_2x - g_3; \qquad E' : y^2 = 4x^3 - m^2g_2x + m^3g_3;$$

$$\text{(15.5b)} \qquad E : y^2 = 4x^3 - g_2x; \qquad E' : y^2 = 4x^3 + 4g_2x.$$

Here (15.5a) concerns the case $K = \mathbf{Q}(\sqrt{-m})$, $0 < m \in \mathbf{Z}$, $\mathfrak{o} \neq \mathbf{Z}[i]$ (even if $g_2g_3 = 0$); (15.5b) applies to the case $\mathfrak{o} = \mathbf{Z}[i]$. If $K = \mathbf{Q}(i)$ and $\mathfrak{o} \neq \mathbf{Z}[i]$, then we use (15.5a) with $m = 1$.

We mention the curves $y^2 = 4x^3 - 4x$ and $y^2 = 4x^3 + 16x$ discussed in §14.19 as easy examples. They are isogenous over $\mathbf{Q}$, isomorphic over $K = \mathbf{Q}(i)$, but not isomorphic over $\mathbf{Q}$. They determine the same zeta function over $\mathbf{Q}$, which equals $L(s, \lambda)$ with the Hecke ideal character λ of §13.7.

(B) We take $\tau \in K \cap H$ and consider a member C_τ of the family of curves $\{C_z\}$ defined in (14.12) with a fixed $\varphi \in \mathscr{M}_1(\mathbf{Q}_{\mathrm{ab}})$ such that $\varphi(\tau) \neq 0$. Put $h = \mathbf{Q}\big(c_2(\tau), c_3(\tau)\big)$ with c_μ as in (14.12). Then C_τ is h-rational and isomorphic to $\mathbf{C}/\mathfrak{a}$ with $\mathfrak{a} = \mathbf{Z}\tau + \mathbf{Z}$. Let c be the cycle on C_τ corresponding to the cycle of $\mathbf{C}/\mathfrak{a}$ represented by the line segment connecting 0 to 1. Now the curve C_τ can be written $C(\tau\omega,\,\omega)$ with $\omega = 2\pi i\varphi(\tau)$ as noted in §14.8. Clearly $\omega = \int_c dx/y$, and so

$$\Omega_{\mathfrak{a}}(C_\tau) = h^\times \mathfrak{o}^\times \cdot 2\pi i\varphi(\tau) \qquad (\mathfrak{a} = \mathbf{Z}\tau + \mathbf{Z}). \tag{15.6}$$

Therefore we can often determine the isomorphism class of C_τ over h by checking the value of $\varphi(\tau)$. We discussed an example with $K = \mathbf{Q}(\sqrt{-7}\,)$ in §14.18.

16. Holomorphic differential operators on the upper half plane

16.1. So far we have taken the weight k to be an integer. But what concerns us in this section is formulas of type (8.7), and so we can take k to be an arbitrary complex number by choosing a suitable branch of $j_\alpha^k = (c_\alpha z + d_\alpha)^k$. We don't have to worry about the ambiguity of such a factor, because once a branch is chosen, then j_α^{k+2p} for $p \in \mathbf{Z}$ can be defined by $j_\alpha^{k+2p} = (c_\alpha z + d_\alpha)^{2p} j_\alpha^k$, and any formula of the type $X(f\|_k\alpha) = (Xf)\|_{k+2p}\alpha$ should be understood with such factors. For instance, we can easily verify that (8.7) is valid in that sense for every $k \in \mathbf{C}$.

We now define a sequence of operators $\{\mathfrak{A}_k^p\}_{p=2}^\infty$ acting on functions f on H by

$$\mathfrak{A}_k^p f = \sum_{\nu=1}^{p} (-1)^{p-\nu} c_\nu^p(k) f^{\nu-1}(f')^{p-\nu} f^{(\nu)}, \tag{16.1}$$

$$c_\nu^p(k) = \binom{p}{\nu} k^{\nu-1}\frac{\Gamma(k+p)}{\Gamma(k+\nu)} \quad \text{if} \quad \nu > 1, \quad c_1^p(k) = (p-1)\frac{\Gamma(k+p)}{\Gamma(k+1)},$$

where we put $f' = (\partial/\partial z)f$ and $f^{(\nu)} = (\partial/\partial z)^\nu f$, though f is merely differentiable and not necessarily analytic; we understand that $\Gamma(k+p)/\Gamma(k+\nu) = \prod_{\mu=\nu}^{p-1}(k+\mu)$ for $0 \le \nu \le p$. The leading term of $\mathfrak{A}_k^p f$ is $k^{p-1}f^{p-1}f^{(p)}$ and the term for $\nu = 1$ is $(-1)^{p-1}(p-1)\prod_{\mu=1}^{p-1}(k+\mu)\cdot(f')^p$. For $2 \le p \le 4$, $\mathfrak{A}_k^p f$ takes the following forms:

$$\mathfrak{A}_k^2 f = kff^{(2)} - (k+1)(f')^2,$$
$$\mathfrak{A}_k^3 f = k^2f^2f^{(3)} - 3k(k+2)ff'f^{(2)} + 2(k+1)(k+2)(f')^3,$$
$$\mathfrak{A}_k^4 f = k^3f^3f^{(4)} - 4k^2(k+3)f^2f'f^{(3)} + 6k(k+2)(k+3)f(f')^2f^{(2)}$$
$$- 3(k+1)(k+2)(k+3)(f')^4.$$

Theorem 16.2. (i) *Let δ_k^ν be the operator defined by (8.5). Then*

$$\mathfrak{A}_k^p f = \sum_{\nu=1}^{p} (-1)^{p-\nu} c_\nu^p(k) f^{\nu-1}(\delta_k f)^{p-\nu} \delta_k^\nu f. \tag{16.2}$$

In other words, $\mathfrak{A}_k^p$ can also be defined by substituting $\delta_k^\nu f$ for $f^{(\nu)}$ in (16.1). Moreover,

$$\mathfrak{A}_k^p(f\|_k\alpha) = (\mathfrak{A}_k^p f)\|_{kp+2p}\,\alpha \quad \textit{for every} \quad \alpha \in SL_2(\mathbf{R}). \tag{16.3}$$

(ii) *If* $f \in \mathscr{N}_k^t(\Phi, \Gamma)$ *with* Φ *and* Γ *as in §8.2, then* $(2\pi i)^{-p}\mathfrak{A}_k^p f$ *belongs to* $\mathscr{N}_{kp+2p}^{tp+p}(\Phi, \Gamma)$. *In particular,* $(2\pi i)^{-p}\mathfrak{A}_k^p f \in \mathscr{S}_{kp+2p}(\Phi, \Gamma)$ *if* $f \in \mathscr{M}_k(\Phi, \Gamma)$.

PROOF. Let $\mathfrak{A}'f$ denote the function on the right-hand side of (16.2). Put $\eta(z) = z - \overline{z} = 2iy$ as we did in (8.3). From (8.11) we see that $\mathfrak{A}'f = \sum_{\nu=0}^{p} \eta^{-\nu} P_\nu f$, where $P_\nu f$ is a polynomial in $f, f', \ldots, f^{(p)}$ with constant coefficients, and we easily see that $P_0 f = \mathfrak{A}_k^p f$. We are going to show that $\mathfrak{A}'f$ is holomorphic if f is holomorphic. Then that will mean that $P_\nu f = 0$ for $\nu > 0$ if f is holomorphic, as η^{-1} is algebraically independent over the field of all meromorphic functions on H. Since $P_\nu f$ is a polynomial in $f, f', \ldots, f^{(p)}$ with constant coefficients, we have $P_\nu f = 0$ for an arbitrary f if $\nu > 0$, which shows that $\mathfrak{A}_k^p = \mathfrak{A}'$, which is (16.2). Thus, hereafter we take f to be holomorphic; then our task is to show that $(\partial/\partial\overline{z})(\mathfrak{A}'f) = 0$. We first note that

$$(16.4) \qquad \eta^2(\partial/\partial\overline{z})\delta_k^m f = m(k+m-1)\delta_k^{m-1} f \quad \text{if} \quad f \text{ is holomorphic},$$

which easily follows from (8.11). We have $\mathfrak{A}'f = \sum_{\nu=1}^{p}(-1)^{p-\nu} c_\nu^p(k) F_\nu$ with $F_\nu = f^{\nu-1}(\delta_k f)^{p-\nu}\delta_k^\nu f$. By (16.4) we have

$$\eta^2 \partial F_\nu/\partial\overline{z} = (p-\nu)kf^\nu(\delta_k f)^{p-\nu-1}\delta_k^\nu f + \nu(k+\nu-1)f^{\nu-1}(\delta_k f)^{p-\nu}\delta_k^{\nu-1} f.$$

The first term on the right-hand side vanishes if $\nu = p$. We can easily verify that $c_\nu^p(k)$ times the second term equals the first term for $\eta^2\partial F_{\nu-1}/\partial\overline{z}$ if $\nu > 2$. If $\nu = 2$, it coincides with $\eta^2\partial F_1/\partial\overline{z}$. Thus, with the factors $(-1)^{p-\nu}$, we obtain the desired equality $(\partial/\partial\overline{z})(\mathfrak{A}'f) = 0$. This proves (16.2).

Once (16.2) is established, (16.3) follows from it by virtue of (8.7). The first part of (ii) can easily be verified by using (16.2), as D_k^ν sends $\mathscr{N}_k^t(\Phi, \Gamma)$ to $\mathscr{N}_{k+2\nu}^{t+\nu}(\Phi, \Gamma)$ by Lemma 8.3 (i). To prove the remaining part, let $f \in \mathscr{M}_k(\Phi, \Gamma)$. By (16.1) and the first part of (ii) we see that $(2\pi i)^{-p}\mathfrak{A}_k^p f$ is a holomorphic element of $\mathscr{N}_{kp+2p}(\Phi, \Gamma)$ which equals $\mathscr{M}_{kp+2p}(\Phi, \Gamma)$. Next, take $\alpha \in G^1$ and put $g(z) = (f\|_k\alpha)(z) = \sum_{n=0}^{\infty} a_n \mathbf{e}(nz/N)$ with $a_n \in \mathbf{C}$ and a suitable $N \in \mathbf{Z}, > 0$. By (16.3), $(\mathfrak{A}_k^p f)\|_{kp+2p}\alpha = \mathfrak{A}_k^p g = \sum_{n=0}^{\infty} b_n \mathbf{e}(nz/N)$ with $b_n \in \mathbf{C}$, and clearly $b_0 = 0$. Thus $g \in \mathscr{S}_{kp+2p}(\Gamma)$, and so $(2\pi i)^{-p}\mathfrak{A}_k^p f \in \mathscr{M}_{kp+2p}(\Phi, \Gamma) \cap \mathscr{S}_{kp+2p}(\Gamma) = \mathscr{S}_{kp+2p}(\Phi, \Gamma)$.

16.3. Let us now consider an operator $\mathfrak{Q}$ acting on functions f on H of the form

$$(16.5) \qquad \mathfrak{Q}f = P(f, f', \ldots, f^{(n)}), \quad f' = (\partial/\partial z)f, \quad \ldots, \quad f^{(n)} = (\partial/\partial z)^n f$$

with a polynomial $P(X_0, X_1, \ldots, X_n)$ in indeterminates X_ν with coefficients in $\mathbf{C}$ satisfying

$$(16.6) \qquad \mathfrak{Q}(f\|_k\alpha) = (\mathfrak{Q}f)\|_m\alpha \quad \text{for every} \quad \alpha \in SL_2(\mathbf{R})$$

with two numbers k and m independent of f and α. We can prove that such an operator can be reduced to a polynomial of $\mathfrak{A}_k^p f$ for $2 \le p \le n$ as described in Theorem 16.4 below, provided $k \neq 0$. For $k = 0$ we need an operator $\mathfrak{B}_0^p$ defined by

$$\mathfrak{B}_0^p f = \mathfrak{A}_2^{p-1}(f') \qquad (3 \le p \in \mathbf{Z}). \tag{16.7}$$

From (16.3) we immediately obtain

$$\mathfrak{B}_0^p(f \circ \alpha) = (\mathfrak{B}_0^p f)\|_{4p-4}\,\alpha \ \text{ for every } \ \alpha \in SL_2(\mathbf{R}). \tag{16.8}$$

Before stating the theorem, we introduce the notion of weight. We say that a polynomial $P(X_0, X_1, \dots, X_n)$ with coefficients in $\mathbf{C}$ is of **weight** κ if it is a $\mathbf{C}$-linear combination of the monomials $X^e = \prod_{\nu=0}^n X_\nu^{e_\nu}$ such that $\sum_{\nu=0}^n \nu e_\nu = \kappa$. For instance, $\mathfrak{A}_k^p f$ as a polynomial in $f, f', \dots, f^{(p)}$ is of weight p; it is also homogeneous of degree p. The polynomial we consider may not involve some of the X_ν. If it is a polynomial $R(X_2, \dots, X_n)$, for example, we still view it as a polynomial in $\{X_\nu\}_{\nu=0}^n$ and speak of the weight as above.

Theorem 16.4. *Suppose an operator $\mathfrak{Q}$ defined by (16.5) satisfies (16.6). Put $P(X_0, X_1, \dots, X_n) = \sum_e c_e X^e$, where $c_e \in \mathbf{C}$ and $X^e = \prod_{\nu=0}^n X_\nu^{e_\nu}$ for $e = (e_1, \dots, e_n)$, $0 \le e_\nu \in \mathbf{Z}$. Denote by P_μ the sum of all $c_e X^e$ with e such that $\sum_{\nu=0}^n e_\nu = \mu$, and put $\mathfrak{Q}_\mu f = P_\mu(f, \dots, f^{(n)})$. Then the following assertions hold:*

(i) $\mathfrak{Q}_\mu(f\|_k\alpha) = (\mathfrak{Q}_\mu f)\|_m\alpha$ *for every* $\alpha \in SL_2(\mathbf{R})$.

(ii) *P_μ is of weight κ with an integer κ such that $m - \mu k = 2\kappa$.*

(iii) *Suppose $k \neq 0$; then $\mathfrak{Q}_\mu f = f^{\mu-\kappa} R(\mathfrak{A}_k^2 f, \dots, \mathfrak{A}_k^n f)$ with a polynomial $R(X_2, \dots, X_n)$ of weight κ. Conversely, given a polynomial $R(X_2, \dots, X_n)$ of weight κ, take a positive integer μ so that $f^{\mu-\kappa} R(\mathfrak{A}_k^2 f, \dots, \mathfrak{A}_k^n f)$ is a polynomial in $f, f', \dots, f^{(n)}$. Then $f \mapsto f^{\mu-\kappa} R(\mathfrak{A}_k^2 f, \dots, \mathfrak{A}_k^n f)$ is an operator of type (16.6) with $m = \mu k + 2\kappa$.*

(iv) *Suppose $k = 0$; let $S(Y_0, Y_3, \dots, Y_n)$ be a polynomial in indeterminates $Y_0, Y_3, \dots, Y_n$ that is a linear combination of monomials $Y_0^{b_0} \prod_{\nu=3}^n Y_\nu^{b_\nu}$ such that*

$$-b_0 + \sum_{\nu=3}^n (\nu - 1) b_\nu = \sigma \tag{16.9}$$

with an integer σ that depends only on S. Put

$$\mathfrak{R} f = (f')^\tau S\big((f')^{-2} f, \mathfrak{B}_0^3 f, \dots, \mathfrak{B}_0^n f\big)$$

with an integer τ such that the right-hand side is a polynomial in $f, f', \dots, f^{(n)}$. Then $\mathfrak{R}$ is an operator of type (16.6) with $k = 0$ and $m = 4\sigma + 2\tau$. Moreover, every $\mathfrak{Q}_\mu$ at the beginning of the theorem with $k = 0$ is of this type with $\sigma = \kappa - \mu$ and $\tau = 2\mu - \kappa$.

PROOF. We can put $\mathfrak{Q} = \sum_{\mu=0}^q \mathfrak{Q}_\mu$ with some q. Since $\mathfrak{Q}_\mu(bf) = b^\mu \mathfrak{Q}_\mu f$ for $b \in \mathbf{C}$, we have

$$\sum_{\mu=0}^{q} b^\mu \mathfrak{Q}_\mu(f\|_k\alpha) = \mathfrak{Q}(bf\|_k\alpha) = \mathfrak{Q}(bf)\|_m\alpha = \sum_{\mu=0}^{q} b^\mu(\mathfrak{Q}_\mu f)\|_m\alpha$$

for every $\alpha \in SL_2(\mathbf{R})$ and every b. Thus we obtain (i). Next, take $\alpha = \mathrm{diag}[t, t^{-1}]$ with $t \in \mathbf{R}^\times$. Then $(f\|_k\alpha)^{(\nu)} = t^{2\nu+k}f^{(\nu)}(t^2z)$, and so from (16.6) we see that $c_e \neq 0$ only if $m = \sum_{\nu=0}^n e_\nu(k+2\nu)$. Therefore, if $\mu = \sum_{\nu=0}^n e_\nu$ and $\kappa = \sum_{\nu=0}^n e_\nu\nu$, then $m = \mu k + 2\kappa$. This proves (ii). To prove (iii), put $g_\nu = f^{(\nu)}/f$ and $A_p = f^{-p}\mathfrak{A}_k^p f$ for any fixed f and $k \neq 0$. Then $f^{-\mu}\mathfrak{Q}_\mu f$ and A_n are polynomials in $g_1, \dots, g_n$ of weight κ and n, respectively. Since the leading term of A_n is $k^{n-1}g_n$, we can express $f^{-\mu}\mathfrak{Q}_\mu f$ as a polynomial in $g_1, \dots, g_{n-1}, A_n$. Replacing g_{n-1} similarly by A_{n-1} and repeating the same procedure successively, we eventually find that $f^{-\mu}\mathfrak{Q}_\mu f = F(g_1, A_2, \dots, A_n)$ with a polynomial F, which is clearly of weight κ. Put $\varphi = f\|_k\alpha$ and $j = cz+d$ with $\alpha = \begin{bmatrix} a & b \\ c & d \end{bmatrix} \in SL_2(\mathbf{R})$. Substitution of φ for f in $f^{-\mu}\mathfrak{Q}_\mu f$ replaces g_1 by $j^{-2}(g_1 \circ \alpha) - kcj^{-1}$ and A_ν by $j^{-2\nu}(A_\nu \circ \alpha)$. Thus

$$\begin{aligned}(f^{-\mu}\mathfrak{Q}_\mu f) \circ \alpha &= j^{m-\mu k}\varphi^{-\mu}\mathfrak{Q}_\mu\varphi \\ &= j^{2\kappa}F\big(j^{-2}(g_1 \circ \alpha) - kcj^{-1},\, j^{-4}(A_2 \circ \alpha),\, \dots,\, j^{-2n}(A_n \circ \alpha)\big) \\ &= F\big(g_1 \circ \alpha - kcj,\, A_2 \circ \alpha,\, \dots,\, A_n \circ \alpha\big),\end{aligned}$$

as F is of weight κ. We see that $j(\alpha^{-1}(z)) = (a-cz)^{-1}$, and so applying α^{-1}, we obtain

$$F(g_1, A_2, \dots, A_n) = F(g_1 - kc(a-cz)^{-1}, A_2, \dots, A_n)$$

for every $\alpha \in SL_2(\mathbf{R})$. Thus clearly F does not involve g_1. This proves the first part of (iii). The converse part follows directly from (16.3).

Next suppose $k = 0$; put $h_\nu = f^{(\nu)}/f'$ and $B_p = (f')^{1-p}\mathfrak{B}_0^p f$. Observe that $(f')^{-\mu}\mathfrak{Q}_\mu f$ and B_p are polynomials in $h_0, h_2, \dots, h_n$. The leading term of B_p is $2^{p-2}h_p$, and so replacing h_p by B_p successively in the same manner as in the proof of (iii), we find that $(f')^{-\mu}\mathfrak{Q}_\mu f = G(h_0, h_2, B_3, \dots, B_n)$ with a polynomial G. Now $G(h_0, h_2, B_3, \dots, B_n)$ is a linear combination of the monomials $h_0^{a_0}h_2^{a_2}B_3^{a_3}\cdots B_n^{a_n}$, and so $\mathfrak{Q}_\mu f$ is a linear combination of

$$f^{a_0}(f')^{a_1}(f'')^{a_2}\prod_{p=3}^{n}(\mathfrak{B}_0^p f)^{a_p}$$

with some $a_1 \in \mathbf{Z}$. (Notice that a_1 may be negative, though the other a_ν are nonnegative.) Now $\mathfrak{B}_0^p f$ is of degree $p-1$ and of weight $2p-2$. Since $\mathfrak{Q}_\mu f$ is of degree μ and weight κ, we have

$$\mu = a_0 + a_1 + a_2 + \sum_{p=3}^{n}(p-1)a_p, \qquad \kappa = a_1 + 2a_2 + \sum_{p=3}^{n}2(p-1)a_p,$$

and so

$$(*)\qquad \kappa - \mu = -a_0 + a_2 + \sum_{p=3}^{n}(p-1)a_p.$$

Put $\varphi = f \circ \alpha$ and $j = cz + d$ with α as above. Substitution of φ for f in $f^{-\mu}\mathfrak{Q}_\mu f$ replaces h_0 by $j^2(h_0 \circ \alpha)$, h_2 by $j^{-2}(h_2 \circ \alpha) - 2cj^{-1}$, and B_p by $j^{2-2p}(B_p \circ \alpha)$, and therefore we have

$$\begin{aligned}&\big[(f')^{-\mu}\mathfrak{Q}_\mu f\big] \circ \alpha = j^{m-2\mu}(\varphi')^{-\mu}\mathfrak{Q}_\mu \varphi\\ &= j^{2\kappa-2\mu}G\big(j^2(h_0 \circ \alpha),\, j^{-2}(h_2 \circ \alpha) - 2cj^{-1},\, j^{-4}(B_3 \circ \alpha),\, \dots,\, j^{2-2n}(B_n \circ \alpha)\big)\\ &= G\big(h_0 \circ \alpha,\, h_2 \circ \alpha - 2cj,\, B_3 \circ \alpha,\, \dots,\, B_n \circ \alpha\big),\end{aligned}$$

in view of $(*)$. Applying α^{-1}, we obtain

$$G(h_0,\, h_2,\, B_3,\, \dots,\, B_n) = G(h_0,\, h_2 - 2c(a - cz)^{-1},\, B_3,\, \dots,\, B_n),$$

and so G does not involve h_2. Thus, removing h_2, we have

$$\begin{aligned}\mathfrak{Q}_\mu f &= (f')^\mu G\big(f/f',\, (f')^{-2}\mathfrak{B}_0^3 f,\, \dots,\, (f')^{1-n}\mathfrak{B}_0^n f\big)\\ &= (f')^{2\mu-\kappa}G\big((f')^{-2}f,\, \mathfrak{B}_0^3 f,\, \dots,\, \mathfrak{B}_0^n f\big)\end{aligned}$$

by virtue of $(*)$. This proves the second part of (iv). The first part follows immediately from (16.8). The proof is now complete.

It should be noted that the exponent $\mu - \kappa$ in (iii) above can be negative. For example, we easily see that

$$(16.10)\qquad f^{-2}\{(k+1)\mathfrak{A}_k^4 f + 3(k+2)(k+3)(\mathfrak{A}_k^2 f)^2\}$$

is a polynomial in $f^{(\nu)}$ for $0 \le \nu \le 4$ of degree 2. Thus $\mu = 2$, $\kappa = 4$, and $m = 2k + 8$ in this case.

16.5. Some results of the same type as Theorem 16.4 were obtained by Rankin [Ran]. He constructed a sequence of operators $\{\psi_n\}_{n=2}^\infty$, where ψ_n sends automorphic forms of weight k to those of weight $kn + 2n$, and showed that all operators of the same type as the above $\mathfrak{Q}$ can be reduced to $\{\psi_n\}$ in the same sense as in Theorem 16.4. However, the set of our forms as a whole is quite different from what he defined. His form ψ_n is defined by a determinant, which has 2^{n-1} terms and involves some awkward coefficients so that his proof requires many different arguments according to the nature of k, whereas our $\mathfrak{A}_k^n f$ has only n terms and simpler coefficients, and $k = 0$ is the only exceptional weight that requires a separate argument. Besides, our proof of (16.3) and that of the above theorem are far shorter than the proof of the corresponding facts for ψ_n in [Ran].

There is a formula of the same nature considered by Henri Cohen. In [C] he introduced the following operator acting on a pair of functions $(f,\, g)$ on H:

$$(16.11)\qquad \mathfrak{F}_{k,m}^p(f,\, g) = \sum_{\nu=0}^{p}(-1)^\nu \binom{p}{\nu} \frac{\Gamma(k+p)\Gamma(m+p)}{\Gamma(k+p-\nu)\Gamma(m+\nu)} f^{(p-\nu)}g^{(\nu)},$$

where the notation for the derivatives is the same as in (16.1). Then he proved

(16.12) $\mathfrak{F}^p_{k,n}(f\|_k\alpha,\, g\|_m\alpha) = \mathfrak{F}^p_{k,n}(f,\, g)\|_{k+m+2p}\,\alpha$ for every $\alpha \in SL_2(\mathbf{R})$.

Let us now present an analogue of (16.2), which will lead to a simpler proof of (16.12).

Theorem 16.6.

(16.13) $$\mathfrak{F}^p_{k,m}(f,\, g) = \sum_{\nu=0}^{p}(-1)^\nu \binom{p}{\nu} \frac{\Gamma(k+p)\Gamma(m+p)}{\Gamma(k+p-\nu)\Gamma(m+\nu)}\delta_k^{p-\nu} f\cdot \delta_m^\nu g.$$

PROOF. This can be proved by the same idea as for (16.2). Let $\mathfrak{R}$ denote the right-hand side of (16.13) and let $\eta = 2iy$ as before. Then $\mathfrak{R} = \sum_{\lambda=0}^{2p}\eta^{-\lambda}\mathfrak{Q}_\lambda$ with polynomials $\mathfrak{Q}_\lambda$ in $f^{(\mu)}$ and $g^{(\nu)}$ with constant coefficients. Clearly $\mathfrak{Q}_0 = \mathfrak{F}^p_{k,m}(f,\, g)$. Assuming f and g to be holomorphic, by (16.4) we have

$$\begin{aligned}&\eta^2(\partial/\partial\overline{z})(\delta_k^{p-\nu} f\cdot\delta_m^\nu g)\\ &= (p-\nu)(k+p-\nu-1)\delta_k^{p-\nu-1} f\cdot\delta_m^\nu g + \nu(m+\nu-1)\delta_k^{p-\nu} f\cdot\delta_m^{\nu-1} g.\end{aligned}$$

Then we easily see that $\eta^2(\partial/\partial\overline{z})\mathfrak{R} = 0$ for holomorphic f and g, and so $\mathfrak{Q}_\lambda = 0$ for $\lambda > 0$ if f and g are holomorphic. Since $\mathfrak{Q}_\lambda$ is a polynomial in $f^{(\mu)}$ and $g^{(\nu)}$ with constant coefficients, we can conclude that $\mathfrak{Q}_\lambda = 0$ for $\lambda > 0$ for arbitrary f and g. Thus $\mathfrak{R} = \mathfrak{Q}_0$ as expected. This completes the proof.

Once (16.13) is established, combining it with (8.7), we immediately obtain (16.12). This is much simpler than the proof of (16.12) given in [C]. However, equalities (16.2) and (16.13) are significant on their own, and should not be viewed as tools for proving (16.3) and (16.12).

16.7. The special case $f = g$ is worth discussing. Define an operator $\mathfrak{C}^p_k$ by

(16.14) $$\mathfrak{C}^p_k f = \sum_{\nu=0}^{p}(-1)^\nu \binom{p}{\nu} \frac{\Gamma(k+p)^2}{\Gamma(k+p-\nu)\Gamma(k+\nu)} f^{(p-\nu)} f^{(\nu)} \quad (p\in 2\mathbf{Z}).$$

This equals $\mathfrak{F}^p_{k,k}(f,\, f)$. We assume p to be even, as the sum vanishes if p is odd. Now, from (16.12) and (16.13) we obtain

(16.15) $\mathfrak{C}^p_k(f\|_k\alpha) = (\mathfrak{C}^p_k f)\|_{2k+2p}\,\alpha$ for every $\alpha \in SL_2(\mathbf{R})$,

(16.16) $$\mathfrak{C}^p_k f = \sum_{\nu=0}^{p}(-1)^\nu \binom{p}{\nu} \frac{\Gamma(k+p)^2}{\Gamma(k+p-\nu)\Gamma(k+\nu)}\delta_k^{p-\nu} f\cdot\delta_k^\nu f.$$

We easily see that $2(k+1)\mathfrak{A}^2_k = \mathfrak{C}^2_k$ and $\mathfrak{C}^4_k f$ equals $2k^{-2}(k+2)(k+3)$ times (16.10). It should be noted that (16.1), (16.7), (16.11), and (16.14) are meanigful for functions defined on an arbitrary open subset of $\mathbf{C}$, and (16.3), (16.8), (16.12), and (16.15) hold for every $\alpha \in SL_2(\mathbf{C})$, provided the symbols involved in the formulas are well defined.

From these we obtain an interesting class of differential equations $\mathfrak{A}_k^p f = 0$ and $\mathfrak{C}_k^p f = 0$. Equalities (16.3) and (16.15) show that the set of solutions in either case is stable under the map $f \mapsto f\|_k\alpha$ for every $\alpha \in SL_2(\mathbf{R})$ or even for $\alpha \in SL_2(\mathbf{C})$. In the case of the equation $\mathfrak{C}_k^p f = 0$ we see that any polynomial $q(z)$ in z of degree $< p/2$ is a solution, and so $q\|_k\alpha$ is also a solution for every $\alpha \in SL_2(\mathbf{C})$. Considering $z^\nu\|_k\alpha$ with $\alpha = \begin{bmatrix} 0 & -c^{-1} \\ c & d \end{bmatrix}$, we eventually find that $\sum_{\nu=0}^{\lambda}(c_\nu z + d_\nu)^{-k-\nu}$ with $\lambda = (p/2) - 1$ and complex constants c_ν, d_ν is a generic solution of $\mathfrak{C}_k^p f = 0$.

As for $\mathfrak{A}_k^p f = 0$, we take $r(z) = z^{-c}$ with $c \in \mathbf{C}$. Then the equation $\mathfrak{A}_k r = 0$ produces a polynomial equation of degree p for c, and $c = 0$ is always a solution. For example, the equation is $c(c-k) = 0$ if $p = 2$ and $c(2c-k)(c-k) = 0$ for $p = 3$. Taking $r\|_k\alpha$, we find that the set of solutions consists of $(az+b)^{-k}$ if $p = 2$, and of $(az+b)^{-k/2}(cz+d)^{-k/2}$ if $p = 3$, where a, b, c, d are complex constants. (In fact, the equation $\mathfrak{A}_k^2 f = 0$ can be solved in an elementary way without using (16.3).) The case $p > 3$ is more complicated. In any case, the nature of the solutions of $\mathfrak{A}_k^p f = 0$ is different from that of the solutions of $\mathfrak{C}_k^p f = 0$, and therefore $\mathfrak{A}_k^p$ and $\mathfrak{C}_k^p$ form different classes of operators. Also, this type of analysis seems to suggest that these operators restricted to $\mathscr{M}_k(\Gamma)$ are injective, which is certainly true for $\mathfrak{A}_k^2$, $\mathfrak{A}_k^3$, and $\mathfrak{C}_k^p$.

We give here some explicit examples of $\mathfrak{A}_k^p f$:

$$
\begin{aligned}
(16.17)\qquad &(2\pi i)^{-2}\mathfrak{A}_4^2 E_4 = (1/60)\Delta, \quad (2\pi i)^{-3}\mathfrak{A}_4^3 E_4 = (-50/7)E_6\Delta, \\
&(2\pi i)^{-2}\mathfrak{A}_6^2 E_6 = (-20/7)E_4\Delta, \quad (2\pi i)^{-2}\mathfrak{A}_{12}^2\Delta = -240E_4\Delta^2, \\
&(2\pi i)^{-2}\mathfrak{A}_1^2 F_1 = 4F_1^3F_3 - 2^{-1}F_3^2,
\end{aligned}
$$

where $F_k = F_k(z;\ c_\gamma/3,\ d_\gamma/3)$ of §10.8. We can find $\mathfrak{A}_k^p f$ also for nearly holomorphic f. For example, from (9.9a, d) and (16.2) we obtain

$$(16.18)\qquad (2\pi i)^{-2}\mathfrak{A}_2^2 E_2 = 4E_2^4 - 10E_2^2E_4 + (7/6)E_2E_6 - (25/12)E_4^2.$$

16.8. We add here a few remarks on the operator L_k defined by

$$(16.19)\qquad L_k f = D_k f + 2kE_2 f,$$

which we mentioned in Lemma 8.3 (iii). We can easily verify that

$$(16.20a)\qquad L_k(f\|_k\alpha) - (L_k f)\|_{k+2}\,\alpha = 2k(E_2 - E_2\|_2\,\alpha)(f\|_k\alpha) \qquad (\alpha \in G_{\mathbf{a}}^1),$$

$$(16.20b)\qquad L_{k+\ell}(fg) = (L_k f)g + f(L_\ell g),$$

$$(16.20c)\qquad L_{km}(f^m) = mf^{m-1}L_k f.$$

From (8.18) we see that $L_{12}\Delta = 0$. Since (16.20c) holds for $m \in \mathbf{R}$, we have $L_k(\Delta^{k/12}) = (k/12)\Delta^{k/12-1}L_{12}\Delta = 0$. Suppose $L_k f = 0$ with a holomorphic function f. Then, by (16.20b), $L_0(\Delta^{-k/12}f) = (L_{-k}\Delta^{-k/12})f + \Delta^{-k/12}L_k f = 0$. Since $L_0 = (2\pi i)^{-1}\partial/\partial z$, this means that $\Delta^{-k/12}f$ is a constant. Therefore every holomorphic solution f of $L_k f = 0$ is a constant times $\Delta^{k/12}$.

APPENDIX

A1. Integration and differentiation under the integral sign

A1.1. Let X and Y be either open subsets of $\mathbf{R}^n$ or $\mathbf{Z}$, or their product. We take the standard discrete topology and the point measure on $\mathbf{Z}$, and so we can speak of $\mathbf{C}$-valued continuous or measurable functions on X and Y. Given a measurable function f on X, we say that f is **integrable** on X if $\int_X |f(x)|dx < \infty$, in which case $\int_X f(x)dx$ is meaningful and finite. If $X \subset \mathbf{Z}$, then $\int_X f(x)dx = \sum_{n \in X} f(n)$. Also for a measurable function $g(x, y)$ for $(x, y) \in X \times Y$, we can speak of its integrability on $X \times Y$. Now we ask the question when

$$\int_X \int_Y g(x, y)dydx = \int_Y \int_X g(x, y)dxdy = \int_{X \times Y} g(x, y)d(x, y) \tag{A1.1}$$

is valid. Here we say that the first double integral is meaningful and finite if $g(x, y)$ is integrable on Y for almost all x and $\int_Y g(x, y)dy$ as a function of x is integrable on X.

Theorem A1.2 (Fubini-Tonelli). *Suppose that one of the following three integrals is meaningful and finite:*

$$\int_X \int_Y |g(x, y)|dydx, \quad \int_Y \int_X |g(x, y)|dxdy, \quad \int_{X \times Y} |g(x, y)|d(x, y).$$

Then the other two and the three integrals of (A1.1) are meaningful and finite, and (A1.1) holds.

This is given in any textbook on real analysis. If $Y = \{n \in \mathbf{Z} \,|\, n > 0\}$ and $X \subset \mathbf{R}^n$, for example, the theorem means that

$$\int_X \sum_{n=1}^{\infty} g_n(x)dx = \sum_{n=1}^{\infty} \int_X g_n(x)dx \tag{A1.2}$$

is justified if $\sum_{n=1}^{\infty} \int_X |g_n(x)|dx < \infty$. In most applications g is continuous, but we cannot assume everything to be continuous for the following reason. Take for example $g(x, y) = x^{-1-\sqrt{|y|}}$ for $X = (1, \infty)$ and $Y = (-1/2, 1/2)$. Then

$\int_X g(x, y)dx < \infty$ or $= \infty$ according as $y \neq 0$ or $y = 0$, so that $h(y) = \int_X g(x, y)dx$ is not continuous in y. However, it is measurable and integrable on Y, and g is integrable on $X \times Y$.

Theorem A1.3. (1) *Let A be a nonempty open subset of* $\mathbf{R}$; *let $f(t, x)$ be a* $\mathbf{C}$*-valued function of $(t, x) \in A \times X$ with X as above such that $|f(t, x)| \leq \varphi(x)$ on $A \times X$ with a function φ integrable over X. Suppose also that f is continuous in t for each fixed x and measurable on X for each fixed t. Put*

$$p(t) = \int_X f(t, x)dx.$$

Then p is continuous.

(2) *Suppose moreover that $(\partial f/\partial t)(t, x)$ exists and is continuous on A, and is measurable on X (in the same sense as above); suppose also that $|(\partial f/\partial t)(t, x)| \leq \psi(x)$ on $A \times X$ with a function ψ integrable over X. Then dp/dt exists and*

$$\frac{dp}{dt} = \int_X \frac{\partial f}{\partial t}(t, x)dx.$$

Proof. Take any sequence $\{a_n\}$ in A converging to a point a of A. The existence of φ allows us to apply the Lebesgue convergence theorem to $\{f(a_n, x)\}$ to find that

$$\lim_{n\to\infty} p(a_n) = \lim_{n\to\infty} \int_X f(a_n, x)dx = \int_X \lim_{n\to\infty} f(a_n, x)dx = \int_X f(a, x)dx = p(a).$$

Since this is valid for any such $\{a_n\}$, p is continuous.

To prove (2), put $q(t) = \int_X (\partial f/\partial t)(t, x)dx$. Our result of (1) shows that q is continuous. Take a small interval $(a - \varepsilon, a + \varepsilon) \subset A$ and for $0 < h < \varepsilon$ observe that

$$\begin{aligned} p(a+h) - p(a) &= \int_X [f(a+h, x) - f(a, x)]dx = \int_X \int_a^{a+h} (\partial f/\partial t)(t, x)dtdx \\ &= \int_a^{a+h} \int_X (\partial f/\partial t)(t, x)dxdt = \int_a^{a+h} q(t)dt. \end{aligned}$$

Here Theorem A1.2 is applicable because of the integrability of ψ. Similarly $p(a) - p(a-h) = \int_{a-h}^{a} q(t)dt$. Therefore $dp/dt = q(t)$.

The differentiability with respect to a complex variable requires weaker assumptions. Indeed we have:

Theorem A1.4. *Let X be as above and let A be a nonempty open connected subset of* $\mathbf{C}$; *let $f(z, x)$ be a* $\mathbf{C}$*-valued continuous function of $(z, x) \in A \times X$, holomorphic in z. Then the following assertions hold:*

(1) *$(\partial/\partial z)^a f(z, x)$ is continuous on $A \times X$ for every $a \in \mathbf{Z}, \geq 0$.*

(2) *Suppose $f(z, x)$ is integrable on X locally uniformly with respect to $z \in A$ in the following sense: for every compact subset K of A there exists a constant M_K such that $\int_X |f(z, x)|dx \leq M_K$ for every $z \in K$. Put*

$$g(z) = \int_X f(z, x)dx.$$

Then g is holomorphic on A. Moreover, $(\partial/\partial z)^a f(z, x)$ is integrable on X locally uniformly with respect to $z \in A$ for every $a > 0$, and

$$\frac{d^a g}{dz^a} = \int_X \frac{\partial^a f}{\partial z^a}(z, x)dx.$$

PROOF. For $c \in A$ and $r > 0$ put $D(c, r) = \{ z \in \mathbf{C} \,|\, |z - c| \le r \}$. Take r so that $D(c, 2r) \subset A$, and denote by γ the circle $|z - c| = 2r$ in the positive direction. Then we have

$$\frac{\partial^a f}{\partial z^a}(z, x) = \frac{a!}{2\pi i}\int_\gamma \frac{f(w, x)}{(w - z)^{a+1}}dw$$

for $z \in D(c, r)$. The integrand $f(w, x)/(w-z)^{a+1}$ is continuous in (w, x, z), and so we obtain the continuity of $\partial^a f/\partial z^a$. By our assumption in (2), $\int_X |f(w, x)|dx \le M$ for every $w \in \gamma$ with a constant M; hence $\int_X |f(w, x)/(w - z)^{a+1}|dx \le Mr^{-a-1}$ if $z \in D(c, r)$, and so $|f(w, x)/(w - z)^{a+1}|$ is integrable on $\gamma \times X$ if $z \in D(c, r)$. Consequently

$$\int_X |(\partial/\partial z)^a f(z, x)|dx \le \frac{a!}{2\pi}\iint_X \left| \frac{f(w, x)}{(w - z)^{a+1}} \right| dx|dw| \le a!2r^{-a}M$$

if $z \in D(c, r)$. Thus $(\partial/\partial z)^a f(z, x)$ is integrable on X locally uniformly with respect to $z \in A$. Now, taking $a = 0$, we see that

$$g(z) = \frac{1}{2\pi i}\int_X \int_\gamma \frac{f(w, x)}{w - z}dw\, dx.$$

If z and z' belong to $D(c, r)$, then

$$|g(z) - g(z')| = \left| \frac{1}{2\pi}\int_X \int_\gamma \left\{ \frac{f(w, x)}{w - z} - \frac{f(w, x)}{w - z'} \right\} dw\, dx \right|$$
$$\le \frac{|z - z'|}{2\pi r^2}\int_X \int_\gamma |f(w, x)|\,|dw|\, dx \le 2r^{-1}M|z - z'|.$$

This shows that g is continuous.

Once the continuity of g is established, for $|z - c| < r$ we have

$$g(z) = \frac{1}{2\pi i}\int_X \int_\gamma \frac{f(w, x)}{w - z}dw\, dx = \frac{1}{2\pi i}\int_\gamma \int_X \frac{f(w, x)}{w - z}dx\, dw = \frac{1}{2\pi i}\int_\gamma \frac{g(w)}{w - z}dw.$$

By a well known principle, this shows that g is holomorphic, and moreover,

$$\frac{d^a g}{dz^a} = \frac{a!}{2\pi i}\int_\gamma \frac{g(w)}{(w - z)^{a+1}}dw = \frac{a!}{2\pi i}\int_\gamma \int_X \frac{f(w, x)}{(w - z)^{a+1}}dx\, dw$$
$$= \frac{a!}{2\pi i}\int_X \int_\gamma \frac{f(w, x)}{(w - z)^{a+1}}dw\, dx = \int_X \frac{\partial^a f}{\partial z^a}(z, x)dx$$

for $|z - c| < r$. This completes the proof.

A2. Fourier series with parameters

A2.1. In the setting of §2.2 we consider a **C**-valued C^∞ function $f(x, y)$ of $(x, y) \in F \times D$, where $F = \mathbf{R}^n/L$ with a lattice L and D is a nonempty open subset of $\mathbf{R}^m$. The Fourier coefficients of f in this case are functions of y given by

$$c_\xi(y) = \mathrm{vol}(F)^{-1}\int_F f(x, y)\mathbf{e}(-{}^t\xi x)dx \qquad (\xi \in \tilde{L}), \tag{A2.1}$$

where $\tilde{L}$ is defined by (2.5). In this case we can show that

$$f(x, y) = \sum_{\alpha\in\tilde{L}} c_\alpha(y)\mathbf{e}({}^t\alpha x), \tag{A2.2}$$

and the right-hand side is absolutely and locally uniformly convergent on $F \times D$. To state this fact in a stronger form, we put $(\partial/\partial x)^a = \prod_{\nu=1}^n(\partial/\partial x_\nu)^{a_\nu}$ for every $a = (a_1, \ldots, a_n) \in \mathbf{Z}^n$ with $a_\nu \geq 0$, and define similarly $(\partial/\partial y)^b$ for every $b \in \mathbf{Z}^m$ with nonnegative components. We also put $|a| = \sum_{\nu=1}^n a_\nu$, $\|\xi\| = \left(\sum_{\nu=1}^n \xi_\nu^2\right)^{1/2}$, and $z^a = \prod_{\nu=1}^n z_\nu^{a_\nu}$ for $z \in \mathbf{C}^n$.

Theorem A2.2. *The Fourier coefficient $c_\xi(y)$ is C^∞ in y, and*

$$(\partial/\partial x)^a(\partial/\partial y)^b f(x, y) = \sum_{\xi\in\tilde{L}}(2\pi i\xi)^a(\partial/\partial y)^b c_\xi(y)\mathbf{e}({}^t\xi x) \tag{A2.3}$$

for every a and b as above, where the right-hand side is absolutely and uniformly convergent on $F \times K$ for every compact subset K of D. In particular, (A2.2) holds.

PROOF. Take $L = \tilde{L} = \mathbf{Z}^n$ for simplicity. If g and h are C^∞ functions on F, then repeated integration by parts shows that

$$\int_F h(x)(\partial/\partial x)^a g(x)dx = (-1)^{|a|}\int_F g(x)(\partial/\partial x)^a h(x)dx.$$

Applying this to $f(x, y)$ and $\mathbf{e}(-{}^t\xi x)$, we obtain

$$\int_F (\partial/\partial x)^a f(x, y)\cdot\mathbf{e}(-{}^t\xi x)dx = (2\pi i\xi)^a c_\xi(y).$$

By (2.7) we have

$$\sum_{\xi\in L}|c_\xi(y)|^2(2\pi\xi)^{2a} = \int_F |(\partial/\partial x)^a f(x, y)|^2 dx. \tag{A2.4}$$

Given a compact subset K of D, we can find a constant $M(a, K)$ such that the right-hand side of (A2.4) is $\leq M(a, K)$ for every $y \in K$. Taking a positive integer $p > n/2$, observe that $\|\xi\|^{2p} = \left(\sum_{\nu=1}^n \xi_\nu^2\right)^p \leq \mu\sum_{|a|=p}\xi^{2a}$ with a constant μ depending only on n and p, so that $\sum_{|a|=p}(2\pi\xi)^{2a} \geq \lambda\|\xi\|^{2p}$ with a constant λ. It is well known that $\sum_{0\neq\xi\in L}\|\xi\|^{-2p}$ is convergent if $2p > n$. Call the sum Z_p and put $M_p = \sum_{|a|=p} M(a, K)$. Then the Schwarz inequality shows that

$$(M_p Z_p)^{1/2} \geq \left\{ \sum_{0 \neq \xi \in L} \left| c_\xi(y) \right|^2 \sum_{|a|=p} (2\pi\xi)^{2a} \right\}^{1/2} \left\{ \sum_{0 \neq \xi \in L} \|\xi\|^{-2p} \right\}^{1/2}$$

$$\geq \sum_{0 \neq \xi \in L} \left| c_\xi(y) \right| \left\{ \sum_{|a|=p} (2\pi\xi)^{2a} \right\}^{1/2} \|\xi\|^{-p} \geq \lambda^{1/2} \sum_{0 \neq \xi \in L} \left| c_\xi(y) \right|$$

if $y \in K$. Thus $\sum_{0 \neq \xi \in L} \left| c_\xi(y) \right|$ is uniformly convergent on K. Applying this to $(\partial/\partial x)^a f(x, y)$, we find that $\sum_{0 \neq \xi \in L} \left| (2\pi\xi)^a c_\xi(y) \right|$ is uniformly convergent on K. By virtue of Theorem A1.3 we obtain (A2.3) for $b = 0$. Now with a fixed b put

$$d_\xi(y) = \int_F (\partial/\partial y)^b f(x, y) \mathbf{e}(-{}^t\xi x) dx.$$

Then Theorem A1.3 shows that $c_\xi(y)$ is C^∞, and $d_\xi(y) = (\partial/\partial y)^b c_\xi(y)$. Therefore, taking $(\partial/\partial y)^b f(x, y)$ in place of $f(x, y)$, we obtain (A2.3) in the most general case.

A3. The confluent hypergeometric function

A3.1. To study the analytic nature of Eisenstein series, we introduce one type of confluent hypergeometric function as follows. For $y > 0$ and $(\alpha, \beta) \in \mathbf{C}^2$ put

(A3.1) $$\tau(y; \alpha, \beta) = \int_0^\infty e^{-yt}(1+t)^{\alpha-1} t^{\beta-1} dt.$$

This is convergent if $\mathrm{Re}(\beta) > 0$, and holomorphic in (α, β) by virtue of the principle of Theorem A1.4. We have obviously

(A3.2) $$(\partial/\partial y)\tau(y; \alpha, \beta) = -\tau(y; \alpha, \beta+1).$$

Since $(1+t)^\alpha = (1+t)^{\alpha-1}(1+t)$, we obtain

(A3.3) $$\tau(y; \alpha+1, \beta) = \tau(y; \alpha, \beta) + \tau(y; \alpha, \beta+1).$$

Integration by parts (taking $(d/dt)\left[e^{-yt}(1+t)^\alpha t^\beta\right]$) shows

(A3.4) $$\beta\tau(y; \alpha+1, \beta) = y\tau(y; \alpha+1, \beta+1) - \alpha\tau(y; \alpha, \beta+1)$$

for $\mathrm{Re}(\beta) > 1$. Let us now put

(A3.5) $$V(y; \alpha, \beta) = e^{-y/2}\, \Gamma(\beta)^{-1}\, y^\beta\, \tau(y; \alpha, \beta).$$

This is called a **Whittaker** (or **confluent hypergeometric**) **function**. From (A3.4) we obtain

(A3.6) $$V(y; \alpha+1, \beta) = V(y; \alpha+1, \beta+1) - \alpha y^{-1} V(y; \alpha, \beta+1).$$

This shows that V can be continued as a holomorphic function in (α, β) to the whole $\mathbf{C}^2$. Now we have

(A3.7) $$y^\beta\tau(y;\,\alpha,\,\beta) = \int_0^\infty e^{-t}(1+y^{-1}t)^{\alpha-1}t^{\beta-1}dt.$$

Therefore we see, at least for $\mathrm{Re}(\beta) > 0$, that

(A3.8) $$\lim_{y\to\infty} e^{y/2}V(y;\,\alpha,\,\beta) = 1.$$

Since this is consistent with (A3.6), we can easily verify that (A3.8) holds uniformly for $(\alpha,\,\beta)$ in any compact subset of $\mathbf{C}^2$.

Lemma A3.2. *For every compact subset K of $\mathbf{C}^2$ there exist two positive constants A and B depending only on K such that*

$$|e^{y/2}V(y;\,\alpha,\,\beta)| \le A\bigl(1+y^{-B}\bigr) \quad \textit{if} \quad (\alpha,\,\beta)\in K.$$

PROOF. If $\mathrm{Re}(\alpha-1) \le n$ with $0 \le n \in \mathbf{Z}$, we have $|(1+t)^{\alpha-1}| \le (1+t)^n = \sum_{k=0}^n \binom{n}{k} t^k$ for $t \ge 0$, and hence for $W(y;\,\alpha,\,\beta) = e^{y/2}V(y;\,\alpha,\,\beta)$ and $\mathrm{Re}(\beta) = b > 0$, we obtain

$$\begin{aligned} |W(y;\,\alpha,\,\beta)| &\le |\Gamma(\beta)|^{-1}y^b\sum_{k=0}^n\binom{n}{k}\int_0^\infty e^{-yt}t^{k+b-1}dt \\ &= \sum_{k=0}^n\binom{n}{k}|\Gamma(\beta)|^{-1}\Gamma(k+b)y^{-k}. \end{aligned}$$

This proves our lemma for $\mathrm{Re}(\beta) > 0$. The general case can be reduced to this special case by (A3.6).

Lemma A3.3. $V(y;\,1-\beta,\,1-\alpha) = V(y;\,\alpha,\,\beta)$.

PROOF. For $\mathrm{Re}(\alpha) > 0$ and $\mathrm{Re}(\beta) > 0$ we have, in view of (2.1h),

$$\begin{aligned} &\Gamma(\beta)\tau(y;\,1-\beta,\,\alpha) = \int_0^\infty e^{-yt}\Gamma(\beta)(1+t)^{-\beta}t^{\alpha-1}dt \\ &= \int_0^\infty e^{-yt}\int_0^\infty e^{-u(1+t)}u^{\beta-1}du\,t^{\alpha-1}dt = \int_0^\infty e^{-u}u^{\beta-1}\int_0^\infty e^{-(u+y)t}t^{\alpha-1}dt\,du \\ &= \Gamma(\alpha)\int_0^\infty e^{-u}(u+y)^{-\alpha}u^{\beta-1}du = \Gamma(\alpha)y^{\beta-\alpha}\tau(y;\,1-\alpha,\,\beta), \end{aligned}$$

which gives the desired equality, as V is holomorphic on the whole $\mathbf{C}^2$.

Lemma A3.4. *If $\mathrm{Re}(\alpha+\beta) > 1$ and $z = x+iy \in H$, then*

$$\sum_{m\in\mathbf{Z}}(z+m)^{-\alpha}(\overline{z}+m)^{-\beta} = i^{\beta-\alpha}(2\pi)^{\alpha+\beta}\sum_{n\in\mathbf{Z}}\mathbf{e}(nx+i|n|y)g_n(y;\,\alpha,\,\beta),$$

where g_n is given by

$$\Gamma(\alpha)\Gamma(\beta)g_n(y;\,\alpha,\,\beta) = \begin{cases} n^{\alpha+\beta-1}\tau(4\pi ny;\,\alpha,\,\beta) & \text{if } n>0, \\ |n|^{\alpha+\beta-1}\tau(4\pi|n|y;\,\beta,\,\alpha) & \text{if } n<0, \\ \Gamma(\alpha+\beta-1)(4\pi y)^{1-\alpha-\beta} & \text{if } n=0. \end{cases}$$

Here, for $v \in \mathbf{C}^\times$ and $\alpha \in \mathbf{C}$ we define v^α by

$$v^\alpha = \exp(\alpha \log(v)), \qquad -\pi < \mathrm{Im}[\log(v)] \leq \pi.$$

Then $v^{\alpha+\beta} = v^\alpha v^\beta$, $v^{m\alpha} = (v^\alpha)^m$ for $m \in \mathbf{Z}$, and $(uv)^\alpha = u^\alpha v^\alpha$ provided $\arg(u)$, $\arg(v)$, and $\arg(u) + \arg(v)$ are all contained in the interval $(-\pi, \pi]$ for suitable choices of $\arg(u)$ and $\arg(v)$.

PROOF. The sum on the left-hand side is absolutely and locally uniformly convergent for $z \in H$ and $\mathrm{Re}(\alpha+\beta) > 1$; the right-hand side minus the constant term is absolutely and locally uniformly convergent for $z \in H$ and all $(\alpha, \beta) \in \mathbf{C}^2$ by virtue of Lemma A3.2. Therefore it is sufficient to prove the formula for $\mathrm{Re}(\alpha) > 0$ and $\mathrm{Re}(\beta) > 0$, since the equality in the general case is obtained by analytic continuation. Put $f(x) = z^{-\alpha}\overline{z}^{-\beta}$ for $z = x + iy$ with a fixed y. By (2.9) we have, for $\mathrm{Re}(\alpha + \beta) > 1$,

$$\sum_{m\in\mathbf{Z}} (z+m)^{-\alpha}(\overline{z}+m)^{-\beta} = \sum_{m\in\mathbf{Z}} f(x+m) = \sum_{n\in\mathbf{Z}} \hat{f}(n)\mathbf{e}(nx),$$

where $\hat{f}(t) = \int_{\mathbf{R}} z^{-\alpha}\overline{z}^{-\beta}\mathbf{e}(-tx)dx$. This is convergent for $\mathrm{Re}(\alpha + \beta) > 1$. Assuming that $\mathrm{Re}(\alpha) > 0$ and $\mathrm{Re}(\beta) > 0$, and employing (2.1h), we have

$$\begin{aligned}
\hat{f}(t) &= i^{\beta-\alpha}\int_{-\infty}^{\infty} (y-ix)^{-\alpha}(y+ix)^{-\beta}\mathbf{e}(-tx)dx \\
&= i^{\beta-\alpha}\Gamma(\alpha)^{-1}\int_{-\infty}^{\infty}\int_0^{\infty} e^{-r(y-ix)}r^{\alpha-1}dr\,(y+ix)^{-\beta}\mathbf{e}(-tx)dx \\
&= i^{\beta-\alpha}\Gamma(\alpha)^{-1}\int_0^{\infty} e^{-ry}r^{\alpha-1}\int_{-\infty}^{\infty} e^{ix(r-2\pi t)}(y+ix)^{-\beta}dx\,dr \\
&= 2\pi\cdot i^{\beta-\alpha}\Gamma(\alpha)^{-1}\Gamma(\beta)^{-1}e^{2\pi ty}\int_{2\pi a}^{\infty} e^{-2ry}r^{\alpha-1}(r-2\pi t)^{\beta-1}\,dr,
\end{aligned}$$

where $a = \mathrm{Max}(0, t)$. Putting $r = 2\pi p$, we thus find that

$$\Gamma(\alpha)\Gamma(\beta)\hat{f}(t) = (2\pi)^{\alpha+\beta}\, i^{\beta-\alpha}\, e^{2\pi ty}\int_a^{\infty} e^{-4\pi py}p^{\alpha-1}(p-t)^{\beta-1}dp.$$

Putting $p - t = tq$ or $p = -tq$ according as $t > 0$ or $t < 0$, we find that the last integral equals $t^{\alpha+\beta-1}e^{-4\pi ty}\tau(4\pi ty, \alpha, \beta)$ or $|t|^{\alpha+\beta-1}\tau(4\pi|t|y, \beta, \alpha)$. If $t = 0$, the integral becomes $\Gamma(\alpha+\beta-1)(4\pi y)^{1-\alpha-\beta}$. This completes the proof.

We now study the analytic nature of the series $\mathfrak{E}_k^N$ of (9.1). We have shown that it is absolutely convergent for $\mathrm{Re}(s) > 1 + k/2$, and therefore it defines a holomorphic function in s on that half-plane. More strongly we have

Theorem A3.5. *There is a real analytic function of $(z, s) \in H \times \mathbf{C}$ which is holomorphic in s and which coincides with $s(s-1)\Gamma(s+k')\mathfrak{E}_k^N(z, s; p, q)$*

for $\mathrm{Re}(2s)+k>2$, *where* $k' = \mathrm{Max}(k, 0)$. *The factor* $s(s-1)$ *is unnecessary if* $k \neq 0$. *If* $k=0$, *then* $\Gamma(s)\mathfrak{E}_0^N(z, s; p, q)$ *has residue* πN^{-2} *at* $s=1$, *and* $-\delta(p/N)\delta(q/N)$ *at* $s=0$, *where* $\delta(x)=1$ *if* $x \in \mathbf{Z}$, *and* $\delta(x)=0$ *otherwise.*

PROOF. We have

$$\mathfrak{E}_k^N(z, s; p, q) = y^s\delta(p/N) \sum_{0\neq n\in q+N\mathbf{Z}} n^{-k}|n|^{-2s}$$
$$+ y^s N^{-k-2s} \sum_{0\neq m\in p+N\mathbf{Z}} \sum_{\ell\in\mathbf{Z}} \left(\frac{mz+q}{N}+\ell\right)^{-k} \left|\frac{mz+q}{N}+\ell\right|^{-2s}.$$

Here the first sum consists of the terms of $\mathfrak{E}_k^N$ with $m=0$. Apply Lemma A3.4 to the sum over ℓ. (If $m<0$, we have $mz+n = -(|m|z-n)$, and so after multiplying by $(-1)^k$, we can employ Lemma A3.4.) Then for $\mathrm{Re}(2s+k)>2$ the second line of the right-hand side can be written

$$(*) \qquad \pi\cdot i^{-k}2^{2-k-2s}N^{-1}y^{1-k-s}\frac{\Gamma(2s+k-1)}{\Gamma(s)\Gamma(s+k)} \sum_{0\neq m\in p+N\mathbf{Z}} m^{-k}|m|^{1-2s}$$
$$+\frac{i^{-k}(2\pi)^{k+2s}N^{-k-2s}y^s}{\Gamma(s)\Gamma(s+k)} \sum_{0\neq t\in\mathbf{Z}} \mathbf{e}\big((tx+i|t|y)/N\big)A_t(y, s),$$

$$A_t(y, s) = \sum_{t=|m|n} \mathbf{e}\big(\varepsilon_m qn/N\big)\varepsilon_m^k|n|^{2s+k-1}\cdot\begin{cases}\tau(4\pi|t|y/N,\ s+k,\ s) & \text{if } t>0,\\ \tau(4\pi|t|y/N,\ s,\ s+k) & \text{if } t<0,\end{cases}$$

where $0\neq n\in\mathbf{Z}$, $0\neq m\in p+N\mathbf{Z}$, and $\varepsilon_m = m/|m|$. Now the magnitude of A_t (appearing in the "nonconstant terms") for an arbitrary $s\in\mathbf{C}$ can be majorized by Lemma A3.2. There are two "constant terms;" one is $y^sM_{q,N}^k(s)$ and the other is

$$\text{(A3.9)} \qquad y^{1-k-s}\frac{\Gamma(2s+k-1)}{\Gamma(s)\Gamma(s+k)}\cdot M_{p,N}^k(s-1/2),$$

disregarding some elementary factors, where $M_{q,N}^k$ is as in (3.5). Their analytic nature can be derived from Theorem 3.4. Thus the whole sum of $(*)$ is meaningful for every $s\in\mathbf{C}$, and we eventutally find the results as stated in our theorem. Though the statements about the poles and residues can be seen from the above expansion, we already proved them in Theorem 9.7.

It should be noted that

$$\text{(A3.10)} \quad \Gamma(s)\Gamma\big(s+(1/2)\big) = \pi^{1/2}\,2^{1-2s}\Gamma(2s), \qquad \Gamma(s)\Gamma(1-s) = \pi/\sin\pi s.$$

A3.6. To study the behavior of $\mathfrak{E}_k^N$ at $s=0$, we first note

$$\text{(A3.11)} \qquad \tau(y;\, n,\, \beta) = \sum_{\mu=0}^{n-1}\binom{n-1}{\mu}\Gamma(\beta+\mu)y^{-\mu-\beta} \qquad (0<n\in\mathbf{Z}),$$

$$\text{(A3.12)}\qquad \left[\frac{\tau(y;\,\alpha,\,\beta)}{\Gamma(\beta)}\right]_{\beta=-m} = \sum_{\mu=0}^{m}\binom{m}{\mu}\prod_{\nu=1}^{\mu}(\nu-\alpha)\cdot y^{m-\mu} \qquad (0 \le m \in \mathbf{Z}).$$

The first formula can be obtained from (A3.1) by applying the binomial theorem to $(1+t)^{n-1}$. From Lemma A3.3 we obtain

$$\frac{\tau(y;\,\alpha,\,\beta)}{\Gamma(\beta)} = \frac{\tau(y;\,1-\beta,\,1-\alpha)}{\Gamma(1-\alpha)}\cdot y^{1-\alpha-\beta}.$$

This combined with (A3.11) proves (A3.12). In particular, taking $n = 1$ and $m = 0$, we obtain

$$\text{(A3.13)}\qquad \tau(y;\,1,\,\beta) = \Gamma(\beta)y^{-\beta},$$

$$\text{(A3.14)}\qquad \big[\tau(y;\,\alpha,\,\beta)/\Gamma(\beta)\big]_{\beta=0} = 1.$$

Take $\beta = 0$ in Lemma A3.4. Then we see that $g_n(y;\,\alpha,\,0) = 0$ if $n \le 0$ and $\mathrm{Re}(\alpha) > 1$, and also that $g_n(y;\,\alpha,\,0) = n^{\alpha-1}/\Gamma(\alpha)$ by (A3.14). Therefore we obtain the Lipschitz formula (2.13) as a special case of Lemma A3.4.

A3.7. Let us now prove (9.4) which concerns the expansion of $\mathfrak{E}_k^N(z;\,p,\,q)$ defined by (9.2) when $k > 0$. We take $s = 0$ in the sum expression in the proof of Theorem A3.5. Naturally we have to evaluate $A_t(y,\,s)/\Gamma(s)$ at $s = 0$. If $t < 0$, the quantity involves $\tau(Y;\,s,\,s+k)/\Gamma(s)$, which is 0 at $s = 0$, as $\tau(Y;\,\alpha,\,k)$ is finite for every $\alpha \in \mathbf{C}$. If $t > 0$, we have $\tau(Y;\,s+k,\,s)/\Gamma(s) = 1$ at $s = 0$ by (A3.14), and obtain the "nonconstant terms" of (9.4) as stated. As for the "constant terms," the only nontrivial point is the value of (A3.9) at $s = 0$. For $k > 2$, $M_{p,N}^k(s-1/2)$ is finite at $s = 0$, and so (A3.9) vanishes. If $k = 1$, we already noted in (9.6a) that $M_{p,N}^1 = 0$ for $p \in N\mathbf{Z}$. If $p \notin N\mathbf{Z}$, then $M_{p,N}^1(-1/2)$ is finite, and we obtain (9.5b) for $k = 1$. Suppose $k = 2$. Then the value in question involves $\Gamma(s)^{-1}M_{p,N}^2(s-1/2)$ at $s = 0$. By (3.6), $M_{p,N}^2(s-1/2) = M_{p,N}^0(s+1/2)$. By Theorem 3.4 the last function has a simple pole at $s = 0$ with residue N^{-1}. (Notice that the factor $(N/\pi)^s\Gamma(s)$ produces $N^{1/2}$ at $s = 1/2$.) Therefore we obtain (9.5b) for $k = 2$. This completes the proof of (9.4).

A3.8. Next we prove (9.14). We assume $k > 1$ and take $s = 1-k$ in the sum expression in the proof of Theorem A3.5. The nonconstant terms involve $\tau(Y;\,s+k,\,s)/\Gamma(s)$ for $t > 0$, which is Y^{k-1} at $s = 0$ by (A3.13). For $t < 0$ we note that $\tau(Y;\,s,\,s+k)$ is finite for $\mathrm{Re}(s+k) > 0$, and so $\tau(Y;\,s,\,s+k)/\Gamma(s) = 0$ for $s = 1-k$. Thus we obtain the nonconstant terms as stated in (9.14). As for the constant terms, the easier one is $y^s\delta(p/N)M_{q,N}^k(s)$. Let $k = 2m-r$ with $m \in \mathbf{Z}$ and $r = 0$ or 1. Then $M_{q,N}^k(s) = D_{q,N}^r(2s+k)$ by (3.12) and $M_{q,N}^k(1-k) = D_{q,N}^r(2-k) = D_{q,N}^r(r+2-2m) = 0$ if $m > 1$ by (3.13a). If $k = 2$ and $m = 1$, then $M_{q,N}^2(-1) = D_{q,N}^0(0) = -\delta(q/N)$ by (3.13b). Therefore we obtain (9.15a). Finally we have to evaluate (A3.9). We have $M_{p,N}^k(s-1/2) = D_{p,N}^r(2s+k-1)$, whose value at $s = 1 - k$ is $D^r_{p,N}(1-k)$, which equals $-2k^{-1}N^{k-1}B_k(p/N)$ by (4.13). To evaluate $\Gamma(2s+k-1)/\Gamma(s)$ at $s = 1-k$, put $m = k-1$ and $s = \sigma - m$. Then

$$\frac{\Gamma(2s+k-1)}{\Gamma(s)} = \frac{\Gamma(2\sigma - m)}{\Gamma(\sigma - m)} = \frac{\Gamma(2\sigma)}{\Gamma(\sigma)} \prod_{\nu=1}^{m} \frac{\sigma - \nu}{2\sigma - \nu},$$

which takes the value $1/2$ at $\sigma = 0$. Combining these together, we obtain (9.15b). This completes the proof of formula (9.14).

A3.9. We end this section by mentioning the paper [S85] which concerns eigenforms of invariant differential operators in the Hilbert modular case, which include generalizations of the Eisenstein series of (9.1). Such forms have Fourier expansions in terms of the functions τ or V. Here we note three noteworthy facts proved in [S85], of which the first two are relevant to the present book: (1) every holomorphic modular form is the sum of a cusp form and an Eisenstein series; (2) the functional equations of such generalized Eisenstein series stated in a form different from Theorem 9.7; (3) it is shown in Proposition 10.1 of that paper that V satisfies a differential equation of the second order.

A4. The Weierstrass $\wp$-function

A4.1. In this section we recall several basic facts on $\wp$. Let $L = \mathbf{Z}\omega_1 + \mathbf{Z}\omega_2$ as in §10.1. We have shown in §7.1 that $\sum_{0\neq\omega\in L} |\omega|^{-s} < \infty$ if $2 < s \in \mathbf{R}$. We now consider

$$f_k(u) = \sum_{\omega\in L} (u-\omega)^{-k} \qquad (u \in \mathbf{C}, \notin L,\ 3 \le k \in \mathbf{Z}).$$

This is uniformly convergent in any compact subset B of $\mathbf{C}$ such that $B \cap L = \emptyset$. Indeed, we can put, for $0 < A \in \mathbf{R}$,

$$f_k(u) = \sum_{|\omega|<2A} (u-\omega)^{-k} + \sum_{|\omega|\ge 2A} (u-\omega)^{-k}.$$

The first sum is a finite sum. If $|\omega| \ge 2A$ and $u \le A$, then $|u-\omega| \ge |\omega|/2$, and so the second sum is majorized by $2^k \sum_{0\neq\omega\in L} |\omega|^{-k}$, which proves the desired fact.

For each $\omega \in L$ we see that $f_k(u) - (u-\omega)^{-k}$ is holomorphic in a neighborhood of ω, and so f_k is a meromorphic function on $\mathbf{C}$ which has a pole of order k at every $\omega \in L$. In particular, take $k = 3$. Then

$$-2f_3(u) = -2u^{-3} - 2\sum_{0\neq\omega\in L} (u-\omega)^{-3}$$

is a meromorphic function on $\mathbf{C}$, and the last infinite series is uniformly convergent on B as above. Therefore

$$u^{-2} - \sum_{0 \neq \omega \in L} \int_0^u \frac{2}{(u-\omega)^3} du$$

is also uniformly convergent on B, if the path of integration is taken within B. Since $-\int_0^u 2(u-\omega)^{-3}du = (u-\omega)^{-2} - \omega^{-2}$, we find that

$$u^{-2} + \sum_{0 \neq \omega \in L} \{(u-\omega)^{-2} - \omega^{-2}\}$$

is a holomorphic function on $\mathbf{C}$ outside of L. We also see that it defines a meromorphic function on $\mathbf{C}$ with a pole of order 2 at every $\omega \in L$. This is called the Weierstrass $\wp$-function and denoted by $\wp(u)$. Strictly speaking, $\wp$ is a function in three varables u, ω_1, ω_2, but for the moment we suppress ω_1 and ω_2,

Let $\wp'(u) = d\wp/du$. Then $\wp'(u) = -2\sum_{\omega \in L}(u-\omega)^{-3}$. We have

$$\wp(u+\omega) = \wp(u) \quad \text{and} \quad \wp'(u+\omega) = \wp'(u) \quad \text{for every } \omega \in L,$$
$$\wp(-u) = \wp(u) \quad \text{and} \quad \wp'(-u) = -\wp'(u).$$

These formulas for $\wp'$ are obvious. Also, changing ω for $-\omega$, we obtain $\wp(-u) = \wp(u)$. Next,

$$(d/du)\big[\wp(u+\omega_\nu) - \wp(u)\big] = \wp'(u+\omega_\nu) - \wp'(u) = 0 \quad \text{for } \nu = 1, 2.$$

Thus $\wp(u+\omega_\nu) - \wp(u) = c_\nu$ with a constant c_ν. Put $u = -\omega_\nu/2$. Then $c_\nu = \wp(\omega_\nu/2) - \wp(\omega_\nu/2) = 0$, which proves that $\wp(u+\omega_\nu) = \wp(u)$ for $\nu = 1, 2$, and consequently $\wp(u+\omega) = \wp(u)$ for every $\omega \in L$.

Our next task is to prove (10.3b) and (10.3d). We start with

$$\wp(u) = u^{-2} + \sum_{0 \neq \omega \in L} \omega^{-2}\{(1-u/\omega)^{-2} - 1\}.$$

Since $(1-x)^{-2} = \sum_{n=1}^{\infty} nx^{n-1}$, we have

$$\wp(u) = u^{-2} + \sum_{0 \neq \omega \in L} \sum_{n=2}^{\infty} nu^{n-1}\omega^{1-n} = u^{-2} + \sum_{n=2}^{\infty} nu^{n-1} \sum_{0 \neq \omega \in L} \omega^{-1-n}.$$

The last sum over ω is nonzero only when $n+1 = 2m$ with $0 < m \in \mathbf{Z}$. Then the sum is G_{2m} of (10.2). Thus $\wp(u) = u^{-2} + \sum_{m=2}^{\infty}(2m-1)G_{2m}u^{2m-2}$, which is (10.3d). From this we find expansions

$$\wp(u) = u^{-2} + 3G_4u^2 + 5G_6u^4 + \cdots,$$
$$\wp'(u) = -2u^{-3} + 6G_4u + 20G_6u^3 + \cdots,$$
$$\wp(u)^3 = u^{-6} + 9G_4u^{-2} + 15G_6 + \cdots,$$
$$\wp'(u)^2 = 4u^{-6} - 24G_4u^{-2} - 80G_6 + \cdots.$$

Put $F(u) = \wp'(u)^2 - 4\wp(u)^3 + 60G_4\wp(u) + 140G_6$. Then we see that F is finite and zero at $u = 0$. Thus F defines a holomorphic function on the comapct Riemann surface $\mathbf{C}/L$, and so it must be a constant. (Take, for example a compact subset K of $\mathbf{C}$ that covers $\mathbf{C}$ modulo L. Let M be the maximum of $|F|$ on K. Then $|F| \leq M$ on the whole $\mathbf{C}$. A bounded holomorphic function on $\mathbf{C}$ must be a constant.) Thus F must be identically equal to 0, as $F(0) = 0$. Putting $g_2 = 60G_4$ and $g_3 = 140G_6$, we obtain (10.3b).

We mention two important facts without proof.

(I) The field $\mathbf{C}(\wp, \wp')$ generated by the functions $\wp$ and $\wp'$ over $\mathbf{C}$ is the field of all meromorphic functions on $\mathbf{C}/L$.

(II) The map $u \mapsto (\wp(u), \wp'(u))$ gives a biregular map of $\mathbf{C}/L$ onto the projective curve associated to the affine curve $Y^2 = 4X^3 - g_2X - g_3$.

A4.2. Here we give a proof of (10.12). We have

$$\wp(u;\, z,\, 1) = u^{-2} + \sum_{0 \neq (m,n) \in \mathbf{Z}^2} \{(u - mz - n)^{-2} - (mz + n)^{-2}\}.$$

Dividing the last sum into the sum for $m = 0$ and the remaining terms, we obtain

$$\begin{aligned}\wp(u;\, z,\, 1) &= u^{-2} + \sum_{0 \neq n \in \mathbf{Z}} \{(u - n)^{-2} - n^{-2}\} \\ &\quad + \sum_{0 \neq m \in \mathbf{Z}} \sum_{n \in \mathbf{Z}} \{(u - mz - n)^{-2} - (mz + n)^{-2}\} \\ &= -2\sum_{n=1}^{\infty} n^{-2} + \sum_{n \in \mathbf{Z}} (u - n)^{-2} + \sum_{m=1}^{\infty} g_m(u,\, z),\end{aligned}$$

where

$$g_m(u,\, z) = \sum_{n \in \mathbf{Z}} \{(u + mz + n)^{-2} + (mz + n - u)^{-2}\} - 2\sum_{n \in \mathbf{Z}} (mz + n)^{-2}.$$

We have $\sum_{n=1}^{\infty} n^{-2} = \pi^2/6$. Also, by (2.14) and (2.16), $\sum_{n \in \mathbf{Z}} (u - n)^{-2} = -4\pi^2 \mathbf{e}(u)\{1 - \mathbf{e}(u)\}^{-2}$ and

$$(2\pi i)^{-2} g_m(u,\, z) = \sum_{n=1}^{\infty} n\{\mathbf{e}(mnz + nu) + \mathbf{e}(mnz - nu)\} - 2\sum_{n=1}^{\infty} n\mathbf{e}(mnz),$$

provided $\mathrm{Im}(z \pm u) > 0$. (The inequality is unnecessary if we use (2.16).) Given $0 < N \in \mathbf{Z}$, $0 \leq r < N$, and $s \in \mathbf{Z}$ such that $(r,\, s) \notin N\mathbf{Z}^2$, put $\zeta = \mathbf{e}(1/N)$, $\mathbf{q} = \mathbf{e}(z)$, and $\mathbf{q}_N = \mathbf{e}(z/N)$; take $u = (rz + s)/N$. Then $\mathbf{e}(u) = \zeta^s \mathbf{q}_N^r$, $\mathbf{e}(mz \pm u) = \zeta^{\pm s} \mathbf{q}_N^{\pm r} \mathbf{q}^m$, and $\mathrm{Im}(z \pm u) > 0$. Therefore we obtain

$$(2\pi i)^{-2} \wp((rz + s)/N;\, z,\, 1) = \frac{1}{12} - 2\sum_{m=1}^{\infty} \sum_{n=1}^{\infty} n\mathbf{q}^{mn} - \frac{\zeta^s \mathbf{q}_N^r}{(1 - \zeta^s \mathbf{q}_N^r)^2}$$

$$+\sum_{m=1}^{\infty}\sum_{n=1}^{\infty} n\mathbf{q}^{mn}(\zeta^{sn}\mathbf{q}_N^{rn}+\zeta^{-sn}\mathbf{q}_N^{-rn}),$$

which gives (10.12).

A5. The action of $G_{\mathbf{A}+}$ on modular forms

A5.1. We now prove the existence of the action of $G_{\mathbf{A}+}$ on $\sum_k \mathscr{A}_k(\mathbf{Q}_{\mathrm{ab}})$ as stated in §14.2. For $\alpha \in G_+$ and a function f on H we put

$$(f|_k\alpha)(z) = (c_\alpha z + d_\alpha)^{-k} f(\alpha(z)). \tag{A5.1}$$

Clearly $f|_k\alpha = \det(\alpha)^{-k/2} f\|_k\alpha$. If $f \in \mathscr{A}_k(\mathbf{Q}_{\mathrm{ab}})$, then $f|_k\alpha \in \mathscr{A}_k(\mathbf{Q}_{\mathrm{ab}})$ by Theorem 1.5. By Theorem 1.9 (ii), $\mathrm{Gal}(\mathbf{Q}_{\mathrm{ab}}/\mathbf{Q})$ acts naturally on $\mathscr{M}_k(\mathbf{Q}_{\mathrm{ab}})$, and this action can be extended to $\sum_k \mathscr{A}_k(\mathbf{Q}_{\mathrm{ab}})$ as noted in §14.2. Thus for $y \in \mathbf{Q}_{\mathbf{A}}^\times$ and $f \in \mathscr{A}_k(\mathbf{Q}_{\mathrm{ab}})$, $f^{\{y\}}$ is a well-defined element of $\mathscr{A}_k(\mathbf{Q}_{\mathrm{ab}})$.

Lemma A5.2. *If $\varphi \in \mathscr{A}_0(\mathbf{Q}_{\mathrm{ab}})$ and $x = \mathrm{diag}[1, t^{-1}]$ with $t \in \prod_p \mathbf{Z}_p^\times$, then $\varphi^{\{t\}}$ coincides with the image of φ under x defined in [S71, §6.6] and written $\varphi^{\tau(x)}$.*

PROOF. By [S71, Proposition 6.1], $\mathscr{A}_0(\mathbf{Q}_{\mathrm{ab}})$ is generated by $J(z)$and $f_a(z)$, where $J(z)$ is as in (10.4d) and

$$f_a(z) = \frac{\mathbf{g}_2(z)\mathbf{g}_3(z)}{\Delta(z)} F_2(z;\, a), \quad a \in \mathbf{Q}^2, \notin \mathbf{Z}^2,$$

with F_2 of (10.10c). Therefore it is sufficient to prove our lemma when φ is J or f_a. This is trivial for J, as $J^{\tau(x)} = J = J^{\{t\}}$. The image of f_a under x is f_{ax} as noted on the last line of [S71, p. 149]. We have shown in (14.3a) that $F_k(z;\, a)^{\{t\}} = F_k(z;\, ax)$. Now $\mathbf{g}_2$, $\mathbf{g}_3$, and Δ are invariant under $\{t\}$. Thus $(f_a)^{\{t\}} = f_{ax} = (f_a)^{\tau(x)}$ as expected. This proves our lemma.

Theorem A5.3. *There is an action of $G_{\mathbf{A}+}$ on $\sum_{k=0}^{\infty} \mathscr{A}_k(\mathbf{Q}_{\mathrm{ab}})$, written $(x, f) \mapsto f^{[x]}$ for $x \in G_{\mathbf{A}+}$ and $f \in \mathscr{A}_k(\mathbf{Q}_{\mathrm{ab}})$, with the following properties:*

(i) $(f+g)^{[x]} = f^{[x]} + g^{[x]}$, $(fg)^{[x]} = f^{[x]}g^{[x]}$.

(ii) $(f^{[x]})^{[y]} = f^{[xy]}$.

(iii) $f^{[\alpha]} = f|\alpha$ *if* $\alpha \in G_+$.

(iv) $f^{[x]} = f^{\{t\}}$ *if* $x = \mathrm{diag}[1, t^{-1}]$ *with* $t \in \prod_p \mathbf{Z}_p^\times$.

(v) $\mathscr{M}_k(\mathbf{Q}_{\mathrm{ab}})$ *are stable under the action of* $[x]$.

(vi) *The action of x on $\mathscr{A}_0(\mathbf{Q}_{\mathrm{ab}})$ is the same as that of x defined in [S71, §6.6].*

(vii) *For every f, the group $\{x \in G_{\mathbf{A}+} \mid f^{[x]} = f\}$ contains U_n for some n.*

PROOF. To simplify our notation, in this proof we write f^x for $f^{[x]}$. For a fixed $k \in \mathbf{Z}$, ≥ 0, put

$$h(z) = \varphi_{1,\chi}^{\ell}(2z)^k, \quad \Gamma = \Gamma^0(4) \cap \Gamma_0(4), \quad \chi(a) = \left(\frac{-1}{a}\right),$$

where $\varphi_{1,\chi}^{\ell}$ is the symbol of (9.10a) defined with $N = 4$. By (9.11d) we have $h\|_k\gamma = \chi(d_\gamma)^k h$ for every $\gamma \in \Gamma$, and from (9.11c, h) we obtain

(1) $$h|_k\iota = i^{-k}h, \qquad \iota = \begin{bmatrix} 0 & -1 \\ 1 & 0 \end{bmatrix}.$$

Put $T = U_4 \cdot \{\mathrm{diag}[1, t] \mid t \in \prod_p \mathbf{Z}_p^\times\}$. This is a subgroup of $G_{\mathbf{A}+}$ and $G_{\mathbf{A}+} = G_+T$ by (14.1d). Given $f \in \mathscr{A}_k(\mathbf{Q}_{\mathrm{ab}})$ and $x \in G_{\mathbf{A}+}$, take $\alpha \in G_+$ and $u \in T$ so that $x = u\alpha$. Since the action of $G_{\mathbf{A}+}$ on $\mathscr{A}_0(\mathbf{Q}_{\mathrm{ab}})$ is already defined in [S71] and $f/h \in \mathscr{A}_0(\mathbf{Q}_{\mathrm{ab}})$, $(f/h)^x$ is a well-defined element of $\mathscr{A}_0(\mathbf{Q}_{\mathrm{ab}})$. We now define f^x by

(2) $$f^x = (f/h)^x \cdot h|_k\alpha.$$

This does not depend on the choice of α and u. Indeed, if $x = w\gamma$ with $\gamma \in G_+$ and $w \in T$, then $\alpha\gamma^{-1} = u^{-1}w \in T \cap G_+ = T \cap G^1 = \Gamma$ and the a-entry of $\alpha\gamma^{-1}$ belongs to $1 + 4\mathbf{Z}$. Therefore $h|_k\alpha = h|_k\gamma$, which proves the desired fact. Taking f to be h, we find that $h^{u\alpha} = h|_k\alpha$, and in particular, $h^u = h$ for every $u \in T$.

From this definition of f^x we can easily derive (i), (iii), and (vi). If x and t are as in (iv), then $f^x = (f/h)^x h$. By Lemma A5.2, $(f/h)^x = (f/h)^{\{t\}} = f^{\{t\}}/h$, and so we obtain (iv). Given $f \in \mathscr{A}_k(\mathbf{Q}_{\mathrm{ab}})$, we can find a positive multiple n of 4 such that $f/h \in \mathscr{A}_0(k_n, \Gamma(n))$. Then f/h is invariant under U_n; see [S71, (6.6.3)]. Thus for $x \in U_n$ we have $f^x = (f/h)h = f$, which proves (vii). Taking X of (14.1d) to be U_n, we see that the images of f under $G_{\mathbf{A}+}$ are of the form $f^\sigma|_k\alpha$ with $\sigma \in \mathrm{Gal}(\mathbf{Q}_{\mathrm{ab}}/\mathbf{Q})$ and $\alpha \in G_+$. From this we obtain (v).

It remains to prove the associativity (ii), whose proof is long; we reduce the problem to simpler cases. We first observe that $f^{uv} = (f^u)^v$, $f^{\alpha\beta} = (f^\alpha)^\beta$, and $f^{u\alpha} = (f^u)^\alpha$ for $\alpha, \beta \in G_+$ and $u, v \in T$. Given $x, y \in G_{\mathbf{A}+}$, put $x = u\alpha$ with $u \in T$ and $\alpha \in G_+$; put also $\alpha y = v\beta$ with $v \in T$ and $\beta \in G_+$. Suppose

(3) $$(g^\alpha)^y = g^{\alpha y} \quad \text{for every} \quad g \in \mathscr{A}_k(\mathbf{Q}_{\mathrm{ab}}).$$

Then $(f^x)^y = (f^{u\alpha})^y = ((f^u)^\alpha)^y = (f^u)^{\alpha y} = (f^u)^{v\beta} = ((f^u)^v)^\beta = (f^{uv})^\beta = f^{uv\beta} = f^{u\alpha y} = f^{xy}$. Therefore the desired (ii) can be reduced to (3). Now suppose (3) is true for some fixed α, y, and h as g. Since $\mathscr{A}_k(\mathbf{Q}_{\mathrm{ab}}) = h\mathscr{A}_0(\mathbf{Q}_{\mathrm{ab}})$ and (ii) is true for $f \in \mathscr{A}_0(\mathbf{Q}_{\mathrm{ab}})$, we see that (3) is true for every $g \in \mathscr{A}_k(\mathbf{Q}_{\mathrm{ab}})$ and for the same α, y. If (3) is true for a fixed α and arbitrary (g, y), then putting $f^\alpha = g$ and taking $\alpha^{-1}y$ as y, we have $g^{\alpha^{-1}y} = (f^\alpha)^{\alpha^{-1}y} = f^{\alpha\alpha^{-1}y} = f^y = (g^{\alpha^{-1}})^y$, which means that (3) is true with α^{-1} in place of α. Therefore it is sufficient to prove

(4) $$(h^\alpha)^y = h^{\alpha y} \quad \text{for every} \quad \alpha \in B \text{ and every } y \in G_{\mathbf{A}+}$$

with a suitable set of generators B of G_+.

Now, given $\alpha \in B$ and $y \in G_{\mathbf{A}+}$, put $y = v\beta$ with $v \in T$ and $\beta \in G^+$ and $\alpha v = u\gamma$ with $u \in T$ and $\gamma \in G_+$. Then $(h^\alpha)^y = (h^\alpha)^{v\beta} = ((h^\alpha)^v)^\beta$ and $h^{\alpha y} = h^{u\gamma\beta} = ((h^u)^\gamma)^\beta = (h^{\alpha v})^\beta$. Thus (4) holds if $h^{\alpha v} = (h^\alpha)^v$. In other words, it is sufficient to prove (4) for $y \in T$.

As for B, recall that $\Gamma(1)$ is generated by ι of (1) and $\begin{bmatrix} 1 & 1 \\ 0 & 1 \end{bmatrix}$. Therefore, by (1.15), G^+ is generated by ι and $P_+ \cap M_2(\mathbf{Z})$. Take $\alpha = \iota$ and $y \in T$. For $c, d \in U$ let us temporarily write $c \equiv d$ if $cd^{-1} \in U_4$. Then $y \equiv \mathrm{diag}[1, t^{-1}]$ with $t \in \prod_p \mathbf{Z}_p^\times$. By (14.1d), $\mathrm{diag}[t^{-1}, t] \in \gamma U_4$ with $\gamma \in G^1$. Then $\gamma \in G^1 \cap \mathrm{diag}[t^{-1}, t]U_4 = \Gamma$. Put $\iota y = u\gamma\iota$ with $u \in G_{\mathbf{A}+}$. Then $u \in U$ and $u \equiv \mathrm{diag}[1, t^{-1}]$. Thus $u \in T$ and $h^{\iota y} = h^{u\gamma\iota} = h^{\gamma\iota} = \chi(d_\gamma)^k h^\iota$. Now $d_\gamma - t \prec 4\mathbf{Z}$, and so $i^{\{t\}} = \chi(d_\gamma)i$. Thus, from (1) we obtain $(h|_k\iota)^y = (i^{-k}h)^y = \chi(d_\gamma)^k i^{-k} h = \chi(d_\gamma)^k h^\iota$, and so $h^{\iota y} = (h^\iota)^y$.

Next let $\alpha = \begin{bmatrix} a & b \\ 0 & d \end{bmatrix} \in P_+ \cap M_2(\mathbf{Z})$. Put $T_m = U_m\{\mathrm{diag}[1, t] \mid t \in \prod_p \mathbf{Z}_p^\times\}$, where $m = 4ad$. Given $y \in T$, we can put, by (14.1d), $y = w\beta$ with $w \in T_m$ and $\beta \in G_+$. Then $w \in \mathrm{diag}[1, t^{-1}]U_m$ with $t \in \prod_p \mathbf{Z}_p^\times$. Take a positive integer s such that $t^{-1} - s \prec m\mathbf{Z}$ and put $\gamma = \begin{bmatrix} a & sb \\ 0 & d \end{bmatrix}$ and $u = \alpha w \gamma^{-1}$. Since $\alpha U_m \alpha^{-1} \subset U_4$, we easily see that $u \in T$. Thus $h^{\alpha y} = h^{\alpha w \beta} = h^{u\gamma\beta} = h^{\gamma\beta}$. On the other hand, we have

$$(5) \qquad (h^\alpha)^y = (h^\alpha)^{w\beta} = (h^\alpha/h)^{w\beta} h^\beta$$

by (2). Put $h(z) = \sum_{r \in 2^{-1}\mathbf{Z}} c_r \mathbf{e}(rz)$. Then

$$h^\alpha(z) = d^{-k} \sum_r c_r \mathbf{e}(rb/d)\mathbf{e}(raz/d), \quad h^\gamma(z) = d^{-k} \sum_r c_r \mathbf{e}(rsb/d)\mathbf{e}(raz/d).$$

These belong to $\mathscr{M}_k(k_m, \Gamma(m))$ with $k_m = \mathbf{Q}(\mathbf{e}(1/m))$. Since $\mathbf{e}(1/m)^{\{t\}} = \mathbf{e}(s/m)$ as noted in (14.4), we have $(h^\alpha)^{\{t\}} = h^\gamma$. Now $h^\alpha/h \in \mathscr{A}_0(k_m, \Gamma(m))$, and so it is invariant under U_m by (14.14a). Therefore $(h^\alpha/h)^w = (h^\alpha)^{\{t\}}/h$ by Lemma A5.2. Combining this with (5), we obtain $(h^\alpha)^y = (h^\gamma/h)^\beta h^\beta = h^{\gamma\beta} = h^{\alpha y}$, which is the desired fact. This completes the proof.

We extend the action of $G_{\mathbf{A}+}$ on $\sum_{k=0}^\infty \mathscr{M}_k(\mathbf{Q}_{\mathrm{ab}})$ to that on $\sum_{k=0}^\infty \mathscr{N}_k(\mathbf{Q}_{\mathrm{ab}})$. We first introduce the following notation:

$$(\mathrm{A}5.2) \qquad \|a\| = [\mathbf{Z} : a\mathbf{Z}] \quad \text{for} \quad a \in \mathbf{Q}_\mathbf{A}^\times.$$

Theorem A5.4. *There is an action of $G_{\mathbf{A}+}$ on $\sum_{k=0}^\infty \mathscr{N}_k(\mathbf{Q}_{\mathrm{ab}})$, written $(x, f) \mapsto f^{[x]}$ for $x \in G_{\mathbf{A}+}$ and $f \in \mathscr{N}_k(\mathbf{Q}_{\mathrm{ab}})$, with the following properties:*

(i) *The restriction of this action to $\mathscr{M}_k(\mathbf{Q}_{\mathrm{ab}})$ is the same as that of Theorem A5.3.*

(ii) *Statements (i), (ii), (iii), and (vii) of Theorem A5.3 are valid for the elements of* $\sum_{k=0}^{\infty} \mathscr{N}_k(\mathbf{Q}_{\mathrm{ab}})$.

(iii) $D_k^p(f^{[x]}) = \|\det(x)\|^p (D_k^p f)^{[x]}$ *if* D_k^p *is the operator of (8.6).*

(iv) *If* $f(z) = \sum_{a=0}^{[k/2]} (2\pi y)^{-a} \sum_{n=0}^{\infty} c_{an}\mathbf{e}(nz/N) \in \mathscr{N}_k(\mathbf{Q}_{\mathrm{ab}})$ *with* $c_{an} \in \mathbf{Q}_{\mathrm{ab}}$ *and* $x = \mathrm{diag}[1, t^{-1}]$ *with* $t \in \prod_p \mathbf{Z}_p^{\times}$, *then*

$$f^{[x]}(z) = \sum_{a=0}^{[k/2]} (2\pi y)^{-a} \sum_{n=0}^{\infty} c_{an}^{\{t\}} \mathbf{e}(nz/N).$$

(v) *Let* $f \in \mathscr{N}_k(\mathbf{Q}_{\mathrm{ab}})$, $g \in \mathscr{A}_k(\mathbf{Q}_{\mathrm{ab}})$, *and* $\tau \in K \cap H$ *with an imaginary quadratic field* K; *also let* $s \in K_{\mathbf{A}}^{\times}$ *and* $r = q(s)^{-1}$ *with* q *as in (14.5). If* $g(\tau)$ *is finite and* $\neq 0$, *then so is* $g^{[r]}(\tau)$ *and*

$$(f/g)(\tau)^{\{s\}} = (f^{[r]}/g^{[r]})(\tau). \tag{A5.3}$$

PROOF. For f as in (iv) and $\sigma \in \mathrm{Gal}(\mathbf{Q}_{\mathrm{ab}}/\mathbf{Q})$ we put formally $f^\sigma = \sum_a (2\pi y)^{-a} \cdot \sum_n c_{an}^\sigma \mathbf{e}(nz/N)$. By Lemma 8.3 (ii), f can be written as a finite sum $f = \sum_a E_2^a h_a$ with $h_a \in \mathscr{M}_{k-2a}(\mathbf{Q}_{\mathrm{ab}})$. We easily see that $E_2^\sigma = E_2$ and $f^\sigma = \sum_a E_2^a h_a^\sigma$. Since $h_a^\sigma \in \mathscr{M}_{k-2a}(\mathbf{Q}_{\mathrm{ab}})$ by Theorem 1.9 (ii), we see that f^σ is an element of $\mathscr{N}_k(\mathbf{Q}_{\mathrm{ab}})$. Given such an f and $x \in G_{\mathbf{A}+}$, we can put $x = u\alpha$ with $u \in U$ and $\alpha \in G_+$. Then we define $f^{[x]}$ by

$$f^{[x]} = f^{[u]}|_k\alpha, \quad f^{[u]} = \sum_a E_2^a h_a^{[u]}. \tag{A5.4}$$

To see that this is independent of the choice of u and α, put $x = v\beta$ with $v \in U$ and $\beta \in G_+$; put also $\gamma = \beta\alpha^{-1}$. Then $\gamma \in \Gamma(1)$, $u = v\gamma$, and $h_a^{[u]} = h_a^{[v]}|_k\gamma$, and so $\left(\sum_a E_2^a h_a^{[u]}\right)|_k\alpha = \left(\sum_a E_2^a h_a^{[v]}\right)|_k\beta$, which gives the desired fact.

Next, let x and t be as in (iv). Then for f as above we have $f^{[x]} = \sum_a E_2^a h_a^{[x]} = \sum_a E_2^a h_a^{\{t\}} = f^{\{t\}}$ as expected. That (i) and (iii) of Theorem A5.3 are valid in the present case is obvious. To prove (vii) of Theorem A5.3 for f as above, take n so that the h_a are all U_n-invariant. Then we have $f^{[u]} = \sum_a E_2^a h_a^{[u]} = f$, which is the desired fact.

To prove (iii), given f and D_k^p, take n so that both f and $D_k^p f$ are U_n-invariant. Given $x \in G_{\mathbf{A}+}$, by (14.1d) we can put $x = u \cdot \mathrm{diag}[1, t^{-1}]\alpha$ with $u \in U_n$, $t \in \prod_p \mathbf{Z}_p^{\times}$, and $\alpha \in G_+$. Then $f^{[x]} = f^{\{t\}}|_k\alpha$ and $(D_k^p f)^{[x]} = (D_k^p f)^{\{t\}}|_{k+2p}\alpha$. We can easily verify, by induction on p, that $(D_k^p f)^\sigma = D_k^p(f^\sigma)$ for every $\sigma \in \mathrm{Gal}(\mathbf{Q}_{\mathrm{ab}}/\mathbf{Q})$. Thus, by (8.7),

$$\begin{aligned}(D_k^p f)^{\{t\}}|_{k+2p}\alpha &= D_k^p(f^{\{t\}})|_{k+2p}\alpha = \det(\alpha)^{-p-k/2} D_k^p(f^{\{t\}})\|_{k+2p}\alpha \\ &= \det(\alpha)^{-p-k/2} D_k^p(f^{\{t\}}\|_k\alpha) = \det(\alpha)^{-p} D_k^p(f^{\{t\}}|_k\alpha).\end{aligned}$$

This proves (iii), as $\det(\alpha) = \|\det(x)\|$. To prove that $(f^{[x]})^{[y]} = f^{[xy]}$, it is sufficient to prove that $(E_2^{[x]})^{[y]} = E_2^{[xy]}$. Now $-24\Delta E_2 = D_{12}\Delta$ by (8.18). Since $(\Delta^{[x]})^{[y]} = \Delta^{[xy]}$, the desired fact follows easily from (iii).

Finally, as for (v), we first prove, for a fixed $\tau \in K \cap H$ and r as in (v),

(A5.5) $$g \in \mathscr{A}_k(\mathbf{Q}_{\mathrm{ab}}) \text{ and } g(\tau) \neq 0, \infty \iff g^{[r]}(\tau) \neq 0, \infty.$$

Indeed, let $\varphi = \Delta^{k/12}$. Then $\varphi \in \mathscr{M}_k(\mathbf{Q})$, and so $\varphi^{[x]}$ for any $x \in G_{\mathbf{A}+}$ is of the form $\varphi|_k\alpha$ with some $\alpha \in G_+$. Thus $\varphi\varphi^{[r]}$ has no zero on H. Therefore, if $g \in \mathscr{A}_k(\mathbf{Q}_{\mathrm{ab}})$ and $g(\tau) \neq 0, \infty$, then $g^{[r]}(\tau)/\varphi^{[r]}(\tau) = (g/\varphi)^{[r]}(\tau) = (g/\varphi)(\tau)^{\{s\}} \neq 0, \infty$ by (14.6), which proves (A5.5). Now, given $\tau \in K \cap H$, take $\alpha \in G_+$ so that $\alpha \in \mathbf{Q}1_2$ and $\alpha(\tau) = \tau$. Then by (12.6a) we have $E_2(\tau) = (\kappa^{-2} - 1)^{-1} g_\alpha(\tau)$, where $\kappa = j_\alpha(\tau)$ and g_α is an element of $\mathscr{M}_2(\mathbf{Q}_{\mathrm{ab}})$ given by (12.3a). With $\varphi = \Delta^{k/12}$ as above, we have

(*) $$E_2(\tau)/\varphi(\tau) = (\kappa^{-2} - 1)^{-1}(g_\alpha/\varphi)(\tau).$$

Let $r = q(s)$ as in (v). Since $\alpha \in q(K^\times)$ by (12.1a), we have $r\alpha = \alpha r$. Thus, applying r to (12.3a), we have $g_\alpha^{[r]} = E^{[r]}\|_2\alpha - E_2^{[r]}$. Evaluating this at $z = \tau$, we obtain $E_2^{[r]}(\tau) = (\kappa^{-2} - 1)^{-1} g_\alpha^{[r]}(\tau)$. Applying $\{s\}$ to (*), by (14.6) we obtain

$$\begin{aligned}\{E_2(\tau)/\varphi(\tau)\}^{\{s\}} &= (\kappa^{-2} - 1)^{-1}(g_\alpha/\varphi)(\tau)^{\{s\}} \\ &= (\kappa^{-2} - 1)^{-1}(g_\alpha^{[r]}/\varphi^{[r]})(\tau) = E_2^{[r]}(\tau)/\varphi^{[r]}(\tau).\end{aligned}$$

Given f and g as in (v), we put $f = \sum_a E_2^a h_a$ as before. Suppose $f \neq 0$; then $k \geq 2$. Put $q_a = \varphi^{-a} g$. Then $q_a \in \mathscr{A}_{k-2a}(\mathbf{Q}_{\mathrm{ab}})$ and

$$(f/g)(\tau) = \textstyle\sum_a (E_2/\varphi)^a(\tau)(h_a/q_a)(\tau).$$

Applying (14.6) to h_a/q_a and employing our result about $(E_2/\varphi)(\tau)^{\{s\}}$, we obtain (v). This completes the proof.

A5.5 We return to the question about the value $E_2(\tau)$ discussed in Section 12. We consider the problem in a somewhat general setting as follows. Let $0 \neq f \in \mathscr{N}_k(\mathbf{Q}, \Gamma(1))$ and $0 \neq g \in \mathscr{A}_k(\mathbf{Q}, \Gamma(1))$. Then f/g is a $\Gamma(1)$-invariant function. Let $\mathfrak{a} = \mathbf{Z}\omega_1 + \mathbf{Z}\omega_2$ be a $\mathbf{Z}$-lattice in K such that $\omega_1/\omega_2 \in H$. We then put

(A5.6) $$f(\mathfrak{a}) = \omega_2^{-k} f(\omega_1/\omega_2).$$

This does not depend on the choice of ω_1, ω_2. Similarly $g(\mathfrak{a})$ can be defined, provided $g(\omega_1/\omega_2)$ is meaningful. For $y \in K_{\mathbf{A}}^\times$ we define the $\mathbf{Z}$-lattice $y\mathfrak{a}$ as in §14.1.

Theorem A5.6. *Let f, g, and $\mathfrak{a}$ be as above, and let $s \in K_{\mathbf{A}}^\times$. If $g(\mathfrak{a})$ is finite and nonzero, then so is $g(s^{-1}\mathfrak{a})$, and*

(A5.7) $$\{f(\mathfrak{a})/g(\mathfrak{a})\}^{\{s\}} = f(s^{-1}\mathfrak{a})/g(s^{-1}\mathfrak{a}).$$

Proof. Let $r = q(s)^{-1}$ with q of (14.5) defined with $\tau = \omega_1/\omega_2$. Put $r = u\alpha$ with $u \in U$ and $\alpha \in G_+$. By Lemma 8.3 (ii) we can put $f = \sum_a E_2^a h_a$ with $h_a \in \mathscr{M}_{k-2a}(\mathbf{Q}, \Gamma(1))$. Then we have $h_a^{[u]} = h_a$ by Lemma 14.3, and so $f^{[r]} = f|_k\alpha$ by (A5.4). Similarly $g^{[r]} = g|_k\alpha$. We have

$$u^{-1}\begin{bmatrix}\omega_1\\ \omega_2\end{bmatrix}s^{-1} = \alpha\begin{bmatrix}\omega_1\\ \omega_2\end{bmatrix} = \begin{bmatrix}\alpha(\tau)\\ 1\end{bmatrix}(c_\alpha\tau + d_\alpha)\omega_2,$$

and so $s^{-1}\mathfrak{a} = \mathbf{Z}\omega_1' + \mathbf{Z}\omega_2'$ with $\omega_1' = \omega_2'\alpha(\tau)$ and $\omega_2' = (c_\alpha\tau + d_\alpha)\omega_2$. Thus $f(s^{-1}\mathfrak{a}) = f\big(\alpha(\tau)\big)(c_\alpha\tau+d_\alpha)^{-k}\omega_2^{-k} = \omega_2^{-k}(f|_k\alpha)(\tau) = \omega_2^{-k}f^{[r]}(\tau)$. We have similarly $g(s^{-1}\mathfrak{a}) = \omega_2^{-k}g^{[r]}(\tau)$. Now $f(\mathfrak{a})/g(\mathfrak{a}) = (f/g)(\tau)$, and so $\{f(\mathfrak{a})/g(\mathfrak{a})\}^{\{s\}}$ $= (f^{[r]}/g^{[r]})(\tau)$ by (A5.3). Therefore we obtain (A5.7).

As examples of f/g, we can mention E_2^2/E_4 and E_2E_4/E_6. It should also be noted that (A5.7) is similar to the behavior of the classical j-values stated in [S71, Theorem 5.7].

References

[B] D. Blasius, On the critical values of Hecke L-series, *Ann. of Math.* 124 (1986), 23–63.

[C] H. Cohen, Sums involving the values at negative integers of L-functions of quadratic fields, *Math. Ann.* 217 (1975), 271–285.

[D] R. M. Damerell, L-functions of elliptic curves with complex multiplication, I, II, *Acta Arith.* 17 (1970), 287–301; 19 (1971), 311–317.

[Ha] H. Hasse, *Über die Klassenzahl Abelscher Zahlkörper,* Akademie-Verlag, Berlin, 1952.

[He27] E. Hecke, Theorie der Eisensteinschen Reihen höhere Stufe und ihre Anwendung auf Funktionentheorie und Arithmetik, *Abh. Math. Sem. Hamburg,* 5 (1927), 199–224 (= *Mathematische Werke,* 461–486).

[He40] E. Hecke, Analytische Arithmetik der positiven Quadratischen Formen, 1940, *Mathematische Werke,* 1959, 789-918.

[Ram] S. Ramanujan, Some properties of Bernoulli's numbers, *J. Indian Math. Soc.* III, 1911, 219–234 (= *Collected Papers,* 1–14).

[Ran] R. A. Rankin, The construction of automorphic forms from the derivatives of a given form, *J. Indian Math. Soc.* 20 (1956), 103–116.

[Ru] R. S. Rumely, A formula for the Grössencharacter of a parametrized elliptic curve, *J. Number Theory,* 17 (1983), 389–402.

[S71] G. Shimura, *Introduction to the arithmetic theory of automorphic functions,* Iwanami Shoten and Princeton Univ. Press, 1971.

[S71b] G. Shimura, On the zeta-function of an abelian variety with complex multiplication, *Ann. of Math.* 94 (1971), 504–533 (= *Collected Papers,* II, 406–435).

[S73] G. Shimura, On modular forms of half integral weight, *Ann. of Math.* 97 (1973), 440–481 (= *Collected Papers,* II, 532–573).

[S75c] G. Shimura, On some arithmetic properties of modular forms of one and several variables, *Ann. of Math.* 102 (1975), 491–515 (= *Collected Papers,* II, 683–707).

[S75d] G. Shimura, On the Fourier coefficients of modular forms of several variables, *Göttingen Nachr. Akad. Wiss. Math.-Phys. Klasse,* 1975, 261–268 (= *Collected Papers,* II, 708–715).

[S76a] G. Shimura, Theta functions with complex multiplication, *Duke Math. J.* 43 (1976), 673–696 (= *Collected Papers,* II, 716–739).

[S76b] G. Shimura, The special values of the zeta functions associated with cusp forms, *Comm. pure appl. Math.* 29 (1976), 783–804 (= *Collected Papers,* II, 740–761).

[S77] G. Shimura, On the periods of modular forms, *Math. Ann.* 229 (1977), 211–221 (= *Collected Papers,* II, 813–823).

[S78a] G. Shimura, On certain reciprocity-laws for theta functions and modular forms, *Acta math.* 141 (1978), 35–71 (= *Collected Papers,* III, 1–37).

[S78c] G. Shimura, The special values of the zeta functions associated with Hilbert modular forms, *Collected Papers,* III, 75–114 (2003). (This is a revised version of the paper that appeared in *Duke Math. J.* 45 (1978), 637–679; corrections, *Duke Math. J.* 48 (1981), 697.)

[S85] G. Shimura, On the Eisenstein series of Hilbert modular groups, *Revista Matemática Iberoamericana,* 1 (1985), 1–42 (= *Collected Papers,* III, 644–685).

[S87] G. Shimura, Nearly holomorphic functions on hermitian symmetric spaces, *Math. Ann.* 278 (1987), 1–28 (= *Collected Papers,* III, 746–773).

[S97] G. Shimura, *Euler Products and Eisenstein series,* CBMS Regional Conference Series in Math. No. 93, Amer. Math. Soc. 1997.

[S98] G. Shimura, *Abelian varieties with complex multiplication and modular functions,* Princeton Univ. Press, 1998.

[S00] G. Shimura, *Arithmeticity in the theory of automorphic forms,* Math. Surv. Monog. vol. 82, Amer. Math. Soc. 2000.

[Web] H. Weber, *Lehrbuch der Algebra* III, Braunschweig 1908.

[Wei] K. Weierstrass, *Formeln und Lehrsätze zum Gebrauche der elliptischen Funktionen,* notes by H. A. Schwarz, Berlin, 1892.

[Y] H. Yoshida, *Absolute CM-periods,* Math. Surv. Monog. vol. 106, Amer. Math. Soc. 2003.

Index